DIAMOND

A Paradox Logic

2nd Edition

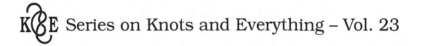

K&E Series on Knots and Everything – Vol. 23

DIAMOND

A Paradox Logic

2nd Edition

N S Hellerstein

City College of San Francisco, USA

 **World Scientific**

NEW JERSEY · LONDON · SINGAPORE · BEIJING · SHANGHAI · HONG KONG · TAIPEI · CHENNAI

Published by

World Scientific Publishing Co. Pte. Ltd.

5 Toh Tuck Link, Singapore 596224

USA office: 27 Warren Street, Suite 401-402, Hackensack, NJ 07601

UK office: 57 Shelton Street, Covent Garden, London WC2H 9HE

Library of Congress Cataloging-in-Publication Data
Hellerstein, N. S. (Nathaniel S.)
    Diamond : a paradox logic / by N.S. Hellerstein. -- 2nd ed.
    p. cm. -- (Series on knots and everything ; v. 23)
    Includes bibliographical references and index.
    ISBN-13: 978-981-4287-13-5 (hardcover : alk. paper)
    ISBN-10: 981-4287-13-X (hardcover : alk. paper)
  1. Logic, Symbolic and mathematical. 2. Paradox. I. Title.

    QA9.H396 2009
    511.3--dc22

                                   2009023583

British Library Cataloguing-in-Publication Data
A catalogue record for this book is available from the British Library.

Printed in Singapore.

Contents

Introduction

There once was a poet from Crete
who performed a remarkable feat
He announced to the wise
"Every Cretan tells lies"
thus ensuring their logic's defeat.

"It cannot be too strongly emphasized that the logical paradoxes are not idle or foolish tricks. They were not included in this volume to make the reader laugh, unless it be at the limitations of logic. The paradoxes are like the fables of La Fontaine which were dressed up to look like innocent stories about fox and grapes, pebbles and frogs. For just as all ethical and moral concepts were skillfully woven into their fabric, so all of logic and mathematics, of philosophy and speculative thought, is interwoven with the fate of these little jokes."

— *Kasner and Newman*, "Paradox Lost and Paradox Regained" from volume 3, "The World of Mathematics"

This book is about "diamond", a logic of paradox. In diamond, a statement can be true yet false; an "imaginary" state, midway between being and non-being. Diamond's imaginary values solve many logical paradoxes unsolvable in two-valued boolean logic.

The purpose of this book is not to bury Paradox but to praise it. I do not intend to explain these absurdities away; instead I want them to blossom to their full mad glory.

I gather these riddles together here to see what they have in common. Maybe they'll reveal some underlying unity, perhaps even a kind of fusion energy! They display many common themes; irony, reverse logic, self-reference, diagonality, nonlinearity, chaos, system failure, tactics versus strategy, and transcendence of former reference frames. Although these paradoxes are truly insoluble as posed, they do in general allow this (fittingly paradoxical!) resolution; namely through loss of resolution! To demand precision is to demand partial vision. These paradoxes define, so to speak, sharp vagueness.

A sense of humor is the best guide to these wild regions. The alternative seems to be a kind of grim defensiveness. There exists a strange tendency for scholars to denigrate these paradoxes by giving them derogatory *names*. Paradoxes have been dubbed "absurd" and "imaginary" and even (O horror!) "irrational". Worse than such bitter insults are the hideously morbid *stories* which the guardians of rationality tell about these agents of Chaos. All too many innocuous riddles have been associated with frightening fables of imprisonment and death; quite gratuitously, I think. It is as if the discoverers of these little jokes hated them and wanted them dead. Did these jests offend some pedant's pride?

Paradox is free. It overthrows the tyranny of logic and thus undermines the logic of tyranny. This book's paradoxes are more subversive than spies, more explosive than bombs, more dangerous than armies, and more trouble than even the President of the United States. They are the weak points in the status quo; they threaten the security of the State. These paradoxes are why the pen is mightier than the sword; a fact which is itself a paradox.

This book has three parts: *Paradox Logic, The Second Paradox,* and *Metamathematical Dilemma.*

Part 1 covers the classic paradoxes of mathematical logic, defines diamond's values and operators, notes diamond's equational laws, introduces diamond's *phase order* lattice, proves that diamond resolves all self-referential systems, resolves the classic paradoxes, demonstrates that diamond embeds the continuum, proves *Zeno's theorem,* graphs *Fuzzy Chaos,* and defines *Clique Theory.*

Part 2 covers diamond as a logic of computation. It defines *sides, star logic,* and *diffraction*; it diffracts *Brownian* and *Kauffman's modulators* to reveal *rotors, pumps,* and *tapes*; these and other forms of *phase computation* can be done on a *Ganglion,* a device here blueprinted.

Part 3 relates diamond to Gödelian metamathematics and non-zero-sum game theory.

This book contains material revised from and added to the previous edition, especially in these sections: Chap. 1, Secs. D, E and H; Chap. 2, Sec. D; Chap. 3, Sec. A; Chap. 5, Sec. B; Chap. 6, Sec. C; Chap. 7, Secs. C, D and G; Chap. 8, Secs. C, D and G; Chap. 9; Chap. 10, Secs. C and G; Chaps. 11 and 12; Chap. 14, Sec. A; Notes and Bibliography.

I would be a liar indeed not to acknowledge my many friends, colleagues and accomplices. These include Louis Kauffman, Dick Shoup, Patrick Grim, Rudy Rucker, Stan Tenen, Tarik Peterson, Sylvia Rippel, Francisco Varela, Douglas Hofstadter, and Raymond Smullyan; without their vital input over many years, this book would have been impossible.

Love and thanks go to my parents, Marjorie and Earl, both of blessed memory; without whom, *I* would have been impossible.

Special thanks go to my dear wife Sherri, without whom I would not have published this.

And thanks to my daughter Hannah, who upon being told the title of Chap. 12, patiently told me, "One...*two*"!

Finally, due credit (and blame!) go to myself, for boldly rushing in where logicians fear to tread.

Said a monk to a man named Joshu
"Is that dog really God"? He said "Mu".
This answer is vexing
And highly perplexing
And that was the best he could do.

Part One

Paradox Logic

Chapter 1
Paradox

A.   The Liar

Epimenides the Cretan said that all Cretans lie; did he tell the truth, or not? Let us assume, for the sake of argument, that every Cretan, except possibly Epimenides himself, was in fact a liar; but what then of Epimenides?

In effect, he says he himself lies; but if he is lying, then he is telling the truth; and if he is telling the truth, then he is lying! Which then is it?

The same conundrum arises from the following sentence:

"This sentence is false".

That sentence is known as the "Liar Paradox", or "pseudomenon".

The pseudomenon obeys this equation:

$$L = \text{not } L.$$

It's true if false, and false if true. Which then is it?

That little jest is King of the Contradictions. They all seem to come back to that persistent riddle. If it is false then it is true, by its own definition; yet if it is true then it is false, for the exact same reason! So which is it, true or false? It seems to undermine dualistic reason itself. Dualists fear this paradox; they would banish it if they could.

Since it is, so to speak, the leader of the Opposition Party, it naturally bears a nasty name; the "Liar" paradox. Don't trust it, say the straight thinkers; and it agrees with them! They denigrate it, but it denigrates *itself*; it admits that it is a liar, and thus it is not *quite* a liar! It is straightforward in its deviation, accurate in its errors, and honest in its lies! Does that make sense to you, dear reader? I must admit that it has never quite made sense to me.

The name "Liar" paradox is nonetheless a gratuitous insult. The pseudomenon merely denies its truth, not its intentions. It may be false innocently, out of lack of ability or information. It may be contradicting itself, not bitterly, as the name "Liar" suggests, but in a milder tone.

Properly speaking, the Liar paradox goes:

"This statement is a lie".

"I am lying".

"I am a liar".

But consider these statements:

"This statement is wrong".

"I am mistaken".

"I am a fool".

This is the Paradox of the Fool; for the Fool is wise if and only if the Fool is foolish! The underlying logic is identical, and rightly so. For whom, after all, does the Liar fool best but the Liar? And whom else does the Fool deceive except the Fool? The Liar is nothing but a Fool, and vice versa!

Therefore I sometimes call the pseudomenon (or Paradox of Self-Denial) the "Fool Paradox", or "Fool's Paradox", or even "Fool's Gold". The mineral "fool's gold" is iron pyrite; a common ore. This fire-y and ironic little riddle is also a common'ore, with a thousand wry offspring.

For instance:

"I am not a Marxist". — *Karl Marx*

"Everything I say is self-serving". — *Richard Nixon*

Tell me, dear reader; would you believe either of these politicos?

Compare the Liar to the following quarrel:

Tweedledee: "Tweedledum is a liar".

Tweedledum: "Tweedledee is a liar".

— *two* calling *each other* liars rather than *one* calling *itself* a liar! This dispute, which I call "Tweedle's Quarrel", is also known as a "toggle".

Its equations are:

$$EE = \text{not } UM$$

$$UM = \text{not } EE$$

This system has two boolean solutions: (true, false) and (false, true). The brothers, though symmetrical, create a difference between them; a memory circuit! It seems that paradox, though chaotic, contains order within it.

Now consider this three-way quarrel:

Moe: "Larry and Curly are liars".

Larry: "Curly and Moe are liars".

Curly: "Moe and Larry are liars".

$$M = \text{not } L \text{ nor } K$$

$$L = \text{not } K \text{ nor } M$$

$$K = \text{not } M \text{ nor } L$$

This system has three solutions: (true, false, false), (false, true, false), and (false, false, true). *One* of the Stooges is honest; but which one?

B. The Anti-Diagonal

Here are two paradoxes of mathematical logic, generated by an "anti-diagonal" process:

Grelling's Paradox. Call an adjective *autological* if it applies to itself, *heterological* if it does not: "*A*" is heterological = "*A*" is not *A*.

Thus, *short* and *polysyllabic* are autological, but *long* and *monosyllabic* are heterological.

> Is *heterological* heterological?
> "Heterological" is heterological = "Heterological" is not
> heterological.
> It is to the extent that it isn't!

Quine's Paradox. Let *quining* be the action of preceding a sentence fragment by its own quotation. For instance, when you quine the fragment *is true when quined*, you get:

> "Is true when quined" is true when quined.

> — a sentence which declares itself true.

In general the sentence:

> "Has property *P* when quined" has property *P* when quined.

is equivalent to the sentence:

> "This sentence has property *P*".

Now consider the sentence:

> "Is false when quined" is false when quined.

That sentence declares itself false. Is it true or false?

C.   Russell's Paradox

Let R be the set of all sets which do not contain themselves:

$$R = \{x \mid x \notin x\}$$

R is an anti-diagonal set. Is it an element of itself?

In general: $x \in R = x \notin x$

and therefore: $R \in R = R \notin R.$

Therefore R is paradoxical. Does R exist?

Here's a close relative of Russell's set; the "Short-Circuit Set":

$$S = \{x : S \notin S\}.$$

S is a constant-valued set, like the universal and null sets:

For all x, $(x \in S) = (S \notin S) = (S \in S).$

All sets are paradox elements for S.

Bertrand Russell told a story about the barber of a Spanish village. Being the only barber in town, he boasted that he shaves all those — and only those — who do not shave themselves. Does the barber shave himself?

To this legend I add a political postscript. That very village is guarded by the watchmen, whose job is to watch all those, and only those, who do not watch themselves. But who shall watch the watchmen?

(Thus honesty in government is truly imaginary!)

The town is also guarded by the trusty watchdog, whose job is to watch all houses, and only those houses, that are not watched by their owners. Does the watchdog watch the doghouse?

Not too long ago that village sent its men off to fight the Great War, which was a war to end all wars, and only those wars, which do not end themselves. Did the Great War end itself?

That village's priest often ponders this theological riddle: God is worshipped by all those, and only those, who do not worship themselves. Does God worship God?

D. Santa and the Grinch

Suppose that a young child were to proclaim:

"If I'm not mistaken, then Santa Claus exists".

If one assumes that Boolean logic applies to this sentence, then its mere existence would imply the existence of Santa Claus!

Why? Well, let the child's statement be symbolized by "R", and the statement "Santa exists" be symbolized by "S". Then we have the equation:

$R = $ if R then $S = R \Rightarrow S = \sim R \vee S$.

Then we have this line of argument:

$R = (R \Rightarrow S)$; assume that R is either true or false.
If R is false, then $R = (\text{false} \Rightarrow S) = (\text{true} \vee S) = \text{true}$.

$R = \text{false}$ implies that $R = \text{true}$;
therefore (by contradiction) R must be true.

Since $R = (R \Rightarrow S)$, $(R \Rightarrow S)$ is also true.
R is true, $(R \Rightarrow S)$ is true; so S is true.
Therefore Santa Claus exists!

This proof uses proof by contradiction; an indirect method, suitable for avoiding overt mention of paradox. Here is another argument, one which confronts the paradox directly:

S is either true or false. If it's true, then so is R:

$$R = (\sim R) \vee \text{ true} = \text{true}.$$

No problem. But if S is false, then R becomes a liar paradox:

$$R = (\sim R) \vee \text{ false} = \sim R.$$

If S is false, then R is non-boolean.

therefore: If R is boolean, then S is true.

Note that both arguments work equally well to prove any other statement besides S to be true; one need merely display the appropriate "santa sentence". Thus, for instance, if some skeptic were to declare:

"If I'm not mistaken, then Santa Claus does not exist".

— then by identical arguments we can prove that Santa Claus does *not* exist!

Given two opposite Santa sentences:

$$R_1 = (R_1 \Rightarrow S); \quad R_2 = (R_2 \Rightarrow \sim S)$$

then at least one of them must be paradoxical.

We can create Santa sentences by Grelling's method. Let us call an adjective "Santa-logical" when it applies to itself only if Santa Claus exists;

"A" is Santa-logical $=$ If "A" is A, then Santa exists.

Is "Santa-logical" Santa-logical?

"Santa-logical" is Santa-logical $=$

     If "Santa-logical" is Santa-logical, then Santa exists.

Here is a Santa sentence via quining:

"Implies that Santa Claus exists when quined" implies that Santa Claus exists when quined.

If that statement is boolean, then Santa Claus exists.

Here's the "Santa Set for sentence G":

$$S_G = \{x \mid (x \in x) \Rightarrow G\}$$

S_G is the set of all sets which contain themselves only if sentence G is true:

$$x \in S_G = ((x \in x) \Rightarrow G).$$

Then "S_G is an element of S_G" equals a Santa sentence for G:

$$S_G \in S_G = ((S_G \in S_G) \Rightarrow G).$$

"$S_G \in S_G$", if boolean, makes G equal true; another one of Santa's gifts. If G is false, then "$S_G \in S_G$" is paradoxical.

One could presumably tell Barber-like stories about Santa sets. For instance, in another Spanish village, the barber takes weekends off; so he shaves all those, and only those, who shave themselves only on the weekend:

B shaves M = If M shaves M, then it's the weekend.

One fine day someone asked: does the barber shave himself?

B shaves B = If B shaves B, then it's the weekend.

Has it been weekends there ever since?

That village is watched by the watchmen, who watch all those, and only those, who watch themselves only when fortune smiles:

W watches C = if C watches C, then fortune smiles.

One fine day someone asked: who watches the watchmen?

W watches W = if W watches W, then fortune smiles.

Does fortune smile on that village?

Recently, that village saw the end of the Cold War, which ended all wars, and only those wars, which end themselves only if money talks:

CW ends W = if W ends W, then money talks.

Did the Cold War end itself?

CW ends CW = if CW ends CW, then money talks.

Does money talk?

That village's priest proclaimed this theological doctrine:

God blesses all those, and only those, who bless themselves only when there is peace:

G blesses S = If S blesses S, then there is peace.

One fine day someone asked the priest: Does God bless God?

G blesses G = If G blesses G, then there is peace.

Is there peace?

Consider the case of Promenides the Cretan, who always disagrees with Epimenides. Recall that Epimenides the Cretan accused all Cretans of being liars, including himself. If we let $E = $ Epimenides, $P = $ Promenides, and $H = $ "honest Cretans exist", then:

$$E = (\sim E) \wedge (\sim H)$$

$$P = \sim E = \sim((\sim E) \wedge (\sim H))$$

$$= E \vee H = (\sim P) \vee H = (P \Rightarrow H)$$

Thus we get this dialog:

Epimenides: All Cretans are liars.

Promenides: You're a liar.

Epimenides: All Cretans are liars, and I am a liar.

Promenides: Either some Cretan is honest, or you're honest.

Epimenides: You're a liar.

Promenides: Either some Cretan is honest, or I'm a liar.

Epimenides: All Cretans are liars, including myself.

Promenides: If I am honest, then some Cretan is honest.

Promenides is the Santa Claus of Crete; for if his statement is boolean, then some honest Cretan exists.

Now suppose that some sarcastic Grinch were to proclaim: "Santa Claus exists, and I am a liar".

$$G = (S \wedge {\sim}G)$$

If Boolean logic applies to this "Grinch Sentence", then it refutes both itself and Santa Claus! For consider this line of argument:

$G = (S \wedge {\sim}G)$; assume that G is either true or false.
If G is true, then $G = (S \wedge {\sim}\mathrm{T}) = \mathrm{F}$.
$G = $ true implies $G = $ false;
   therefore (by contradiction) G must be false.
False $= G = (S \wedge {\sim}G) = (S \wedge {\sim}\mathrm{F}) = S$.

Therefore S is false. Therefore Santa Claus does not exist!

This proof uses proof by contradiction; an indirect method, suitable for avoiding overt mention of paradox. Here is another argument, one which confronts the paradox directly:

S is either true or false. If it's false, then so is G:

$$G = (F \wedge {\sim}G) = \text{false.}$$

No problem. But if S is true, then G becomes a liar paradox:

$$G = (T \wedge {\sim}G) = {\sim}G.$$

If S is true, then G is non-boolean.
Therefore; if G is boolean, then S is false.   QED

Call an adjective *Grinchian* if and only if it does not apply to itself, and Santa Claus exists:

"*A*" is Grinchian = Santa exists, and "*A*" is not *A*.

Is "Grinchian" Grinchian?

"*G*" is *G* = Santa exists, and "*G*" is not *G*.

The "Grinch Set for sentence *H*" is:

$$G_H = \{x \mid H \wedge (x \notin x)\}$$

$$(G_H \in G_H) = (H \wedge (G_H \notin G_H))$$

In paradox logic, the threatened paradox need not affect any other truth value. If Santa Claus *does* exist after all, then the Grinch is exposed as a Liar!

The Grinch sets suggest Grinch stories. Consider the Weekend Barber, who only shaves on the weekends, and only those who do not shave themselves:

WB shaves *M* = It's the weekend, and *M* does not shave *M*.

Does the Weekend Barber shave himself?

WB shaves *WB* = It's the weekend, and *WB* does not shave *WB*.

Note that Epimenides's statement:

"All Cretans are liars, including myself".

— makes him the Grinch of Crete!

E. Antistrephon

That is, "The Retort". This is a tale of the law-courts, dating back to Ancient Greece. Protagoras agreed to train Euathius to be a lawyer, on the condition that his fee be paid, or not paid, according as Euathius win, or lose, his first case in court. (That way Protagoras had an incentive to train his pupil well; but it seems that he trained him too well!) Euathius delayed starting his practice so long that Protagoras lost patience and brought him to court, suing him for the fee. Euathius chose to be his own lawyer; this was his first case.

Protagoras said, "If I win this case, then according to the judgement of the court, Euathius must pay me; if I lose this case, then according to our contract he must pay me. In either case he must pay me".

Euathius retorted, "If Protagoras loses this case, then according to the judgement of the court I need not pay him; if he wins, then according to our contract I need not pay him. In either case I need not pay".

How should the judge rule?

Here's another way to present this paradox:

According to the contract, Euathius will avoid paying the fee — that is, win this lawsuit — exactly if he loses his first case; and Protagoras will get the fee — that is, win this lawsuit — exactly if Euathius wins his first case. But this lawsuit *is* Euathius's first case, and he will win it exactly if Protagoras loses. Therefore Euathius wins the suit if and only if he loses it; ditto for Protagoras.

F. Parity of Infinity

What is the *parity* of infinity? Is infinity odd or even? In the standard Cantorian theory of infinity, ∞ equals its own successor:

$$\infty = \infty + 1$$

But therefore ∞ is even if and only if it is odd! Since ∞ is a *counting* number, it is presumably an integer; but any integer is even *or else* odd!

Infinity has paradoxical parity. We encounter this paradox when we try to define the limit of an infinite oscillation. Consider the sequence $\{x_0, x_1, x_2, \ldots\}$:

$x_0 = \text{True};$

$x_{n+1} = \sim x_n, \quad \text{for all } n.$

Therefore $x_{\text{even}} = \text{True}$, and $x_{\text{odd}} = \text{False}$.

The x's make an oscillation: $\{T, F, T, F, \ldots\}$ Now, can we define a *limit* of this sequence? $\text{Lim}(x_n)$ equals what? If we cannot define this limit, in what sense does infinity have a parity at all? And if no parity, why other arithmetical properties?

We can illuminate the Parity of Infinity paradox with a fictional lamp; the Thompson Lamp, capable of infinitely making many power-toggles in a finite time. The Thompson Lamp clicks on for one minute, then off for a half-minute, then back on for a quarter-minute; then off for an eighth-minute; and so on, in geometrically decreasing intervals until the limit at two minutes, at which point the Lamp stops clicking. After the second minute, is the Lamp on or off?

G.   The Heap

Surely one sand grain does not make a heap of sand. Surely adding another grain will not make it a heap. Nor will adding another, or another, or another. In fact, it seems absurd to say that adding one single grain of sand will turn a non-heap into a heap. By adding enough ones, we can reach any finite number; therefore no finite number of grains of sand will form a sand heap. Yet sand heaps exist; and they contain a finite number of grains of sand!

Let's take it in the opposite direction. Let us grant that a finite sand heap exists. Surely removing one grain of sand will not make it a non-heap. Nor will removing another, nor another, nor another. By subtracting enough ones, we can reduce any finite number to one. Therefore one grain of sand makes a heap!

What went wrong?

Let's try a third time. Grant that one grain of sand forms no heap; but that some finite number of grains do form a heap. If we move a single grain at a time from the heap to the non-heap, then they will eventually become indistinguishable in size. Which then will be the heap, and which the nonheap?

The First Boring Number. This is closely related to the paradox of the Heap. For let us ask the question: are there any boring (that is, uninteresting) numbers? If there are, then surely that collection has a *smallest* element; the *first* uninteresting number. How interesting!

Thus we find a contradiction; and this seems to imply that there are no uninteresting numbers!

But in practice, most persons will agree that most numbers are stiflingly boring, with no interesting features whatsoever! What then becomes of the above argument?

Simply this; that the *smallest* boring number is inherently paradoxical. If being the first boring number were a number's only claim to our interest, then we would find it interesting if and only if we do *not* find it interesting.

Which then is it?

Berry's Paradox. What is "the smallest number that cannot be defined in less than 20 syllables"? If this defines a number, then we have done so in 19 syllables! So this defines a number if and only if it does not.

Presumably Berry's number equals the first boring number, if your boredom threshold is 20 syllables.

These paradoxes connect to the paradox of the Heap by simple psychology. If, for some mad reason, you actually *did* try to count the number of grains in a sand heap, then you will eventually get bored with such an absurd task. Your attention would wander; you would lose track of all those sand grains; errors would accumulate, and the number would become indefinite.

The Heap arises at the onset of uncertainty. In practice, the Heap contains a boring number of sand grains; and the smallest Heap contains the smallest boring number of sand grains!

H. Finitude

Finite is the opposite of infinite; but in paradox-land, that's no excuse! In fact the concept of finiteness is highly paradoxical; for though finite numbers are finite individually and in finite groups, yet they form an infinity.

Let us attempt to *evaluate* finiteness. Let F = 'Finitude', or 'Finity'; the *generic* finite expression. You may replace it with any finite expression.

Is Finity finite?

If F is finite, then you can replace it by $F+1$, and thus by $F+2$, $F+3$, etc. But such a substitution, indefinitely prolonged, yields an infinity.

If F is not finite, then you may not replace F by F, nor by any expression involving F; you must replace F by a well-founded finite expression, which will then be limited.

Therefore F is finite if and only if it is not finite.

Finitude is *just short* of infinity! It is infinity seen from underneath. You may think of it as that mysterious 'large finite number' N, bigger than any number you care to mention; that is, bigger than any *interesting* number.

Call a number "large" if it is bigger than any number you care to mention; that is, bigger than any interesting number. Call a number "medium" if it is bigger than some boring number but less than some interesting number. Call a number "small" if it is less than any boring number.

Then Finitude is the smallest large number; that is, the smallest number bigger than any interesting number. How interesting!

Finitude is dual to the Heap, which is the largest number less than any uninteresting number. The Heap is the lower limit of boredom; Finitude is the upper limit of interest.

We have these inequalities:

small interesting numbers

    $<$ The Heap $=$ first boring number $=$ last small number

    $<$ medium numbers

    $<$ Finitude $=$ last interesting number $=$ first large number

    $<$ large boring numbers

Consider this Berry-like definition:

"One plus the largest number defineable in less than 20 syllables".

If this defines a number, then it has done so in only 19 syllables, and therefore is its own successor. If your boredom threshold is 20 syllables, then this number $=$ "one plus the last interesting number" $=$ Finitude.

Now consider "well-foundedness". Set theorists were so disturbed by Russell's paradox that they decided to acknowledge only "well-founded" sets. A set is well-founded if and only if it has no "infinite descending element chains"; that is, there is no infinite sequence of sets $X_1, X_2, X_3, \ldots$ such that

$$\cdots X_4 \in X_3 \in X_2 \in X_1 \in X_0.$$

Well-founded sets include $\{\ \}$; $\{\{\ \}\}$; $\{\{\{\}\}, \{\{\},\{\{\}\}\}\}$; and even infinite sets such as $\{\{\},\{\{\}\}, \{\{\{\}\}\}, \{\{\{\{\}\}\}\},\ldots\}$; for well-founded sets can be infinitely "wide", so long as they are finitely "deep" along each "branch".

On the other hand, well-foundedness excludes sets such as

$$A = \{A\} = \{\{\{\{\{\{\ldots\}\}\}\}\}\}$$

for it has the infinite descending element chain $\cdots \in A \in A \in A$.

Let *WF* be the set containing all well-founded sets. Is *WF* well-founded?

If *WF* *is* well-founded, then *WF* is in *WF*; but this yields the infinite descending element chain $\cdots \in WF \in WF \in WF$.

On the other hand, if *WF* is *not* well-founded, then any element of *WF* is well-founded, and element chains deriving from those will be finite. Thus all element chains from *WF* will be finite; and therefore *WF* would be well-founded.

And so we see that the concept of "well-foundedness" leads us to the paradox of Finitude.

Related to "well-foundedness" is "clear-foundedness". Call a set "clear-founded" if none of its descending chains are more than a Heap of sets deep. A set is clear-founded if it has only interesting depth. Is the class of clear-founded sets clear-founded?

And just as infinity has paradoxical parity, so do Finitude and the Heap. Is the first boring number odd or even? Is the last interesting number?

I.   Game Paradoxes

Hypergame and the Mortal

Let "Hypergame" be the game whose initial position is the set of all "short" games — that is, all games that end in a finite number of moves. For one's first move in Hypergame, one may move to the initial position of any short game.

Is Hypergame short?

If Hypergame is short, then the first move in Hypergame can be too — Hypergame! But this implies an endless loop, thus making Hypergame no longer a short game!

But if Hypergame is *not* short, then its first move must be into a short game; thus play is bound to be finite, and Hypergame a short game.

The Hypergame paradox resembles the paradox of Finitude. Presumably Hypergame lasts Finitude moves; one plus the largest number definable in less than twenty syllables.

Dear reader, allow me to dramatize this paradox by means of a fictional story about a mythical being. This entity I shall dub "the Mortal"; an unborn spirit who must now make this fatal choice; to choose some mortal form to incarnate as, and thus be be doomed to certain death.

The Mortal has a choice of dooms. Is the Mortal doomed?

Normalcy and the Rebels

Define a game as "normal" if and only if it does not offer the option of moving to its own starting position:

G is normal = the move $G \Rightarrow G$ is not legal.

Let "Normalcy" be the game of all normal games. In it one can move to the initial position of any normal game:

The move $N \Rightarrow G$ is legal = the move $G \Rightarrow G$ is not legal.

Is Normalcy normal? Let $G = N$:

The move $N \Rightarrow N$ is legal = the move $N \Rightarrow N$ is not legal.

Normalcy is normal if and only if it is *ab*normal!

That was Russell's paradox for game theory. Now consider this:

The Rebel is a being who must become one who changes. The Rebel may become all those, and only those, who do not remain themselves:

R may become $B = B$ may not become B.

Can the Rebel remain a Rebel?

R may become $R = R$ may not become R.

A Santa Rebel may become all those, and only those, who remain themselves only if Santa Claus exists:

SR may become $B = ((B$ may become $B) \Rightarrow$ Santa exists)

Therefore:

SR may become $SR = ((SR$ may become $SR) \Rightarrow$ Santa exists)

If the pivot bit is boolean, then Santa Claus exists!

J.   Cantor's Paradox

Cantor's proof of the "uncountability" of the continuum relies on an "anti-diagonalization" process. Suppose we had a countable list of the real numbers between 0 and 1:

$$R_1 = 0.D_{11} D_{12} D_{13} D_{14} \ldots$$

$$R_2 = 0.D_{21} D_{22} D_{23} D_{24} \ldots$$

$$R_3 = 0.D_{31} D_{32} D_{33} D_{34} \ldots$$

$$R_4 = 0.D_{41} D_{42} D_{43} D_{44} \ldots$$

$$\vdots$$

where D_{NM} is the Mth binary digit of the Nth number. Then we define Cantor's "anti-diagonal" number:

$$C = 0.{\sim}D_{11}, {\sim}D_{22}, {\sim}D_{33}, {\sim}D_{44}, \ldots$$

If $C = R_N$ for any N, then $D_{NX} = {\sim}D_{XX}$;

Therefore $D_{NN} = {\sim}D_{NN}$; the pivot bit buzzes.

From this paradox, Cantor deduced that the continuum has too many points to be counted, and thus is of a "higher order" of infinity. Thus a single buzzing bit implies infinities beyond infinities! Was more ever made from less?

I say, why seek "transfinite cardinals", whatever those are? Why not ask for Santa Claus? In this spirit, I introduce the *Santa*-diagonal number:

$$S = 0.(D_{11} \Rightarrow \text{Santa}), (D_{22} \Rightarrow \text{Santa}), (D_{33} \Rightarrow \text{Santa}) \cdots$$

If $S = R_M$ for any M, then $D_{MX} = (D_{XX} \Rightarrow \text{Santa exists})$;

Therefore $D_{MM} = (D_{MM} \Rightarrow \text{Santa exists})$.

If the pivot bit is boolean, then Santa Claus exists!

K.   Paradox of the Boundary

The continuum is paradoxical because it is *continuous*, and boolean logic is discontinuous. This topological difference yields a logical riddle which I call the Paradox of the Boundary.

The paradox of the boundary has many formulations, such as:

What day is midnight?

Is noon A.M. or P.M.?

Is dawn day or night? Is dusk?

Which hemisphere is the equator on?

Which longitude are the poles at?

Which country owns the border?

Is zero plus or minus?

If a statement is true at point A and false at point B, then somewhere in-between lies a boundary. At any point on the boundary, is the statement true, or is it false?

(If line segment AB spanned the island of Crete, then somewhere in the middle we should, of course, find Epimenides!)

Chapter 2

Diamond

A. The Buzzer

Chapter 1 posed the problem of paradox but left it undecided. Is the Liar true or false? Boolean logic cannot answer. What bold expedient would *decide* the question?

What but Experiment? Let us be scientific! Is it possible to build a physical model of formal paradox using simple household items, such as (say) wires, switches, batteries and relays?

Yes, you can! And indeed it's simple! It's *easy*! Just wire together a spring-loaded relay, a switch, and a battery, using this childishly simple circuit:

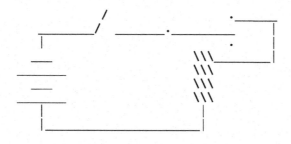

When you close the switch, the relay is caught in a dilemma; for if current flows in the circuit, then the relay shall be energized to break the circuit, and current will stop; whereas if there is no current in the circuit, then the spring-loaded circuit will re-connect, and current will flow. Therefore the relay is open if and only if it is closed. Which?

To find out, close the switch. What do you see?

You would see less than you'd hear; for in fact you would see a blur! The relay would oscillate. It would vibrate. It would, in fact, *buzz*!

All buzzers, bells, alternators, and oscillators are based on this principle of oscillation via negative feedback. Thermostats rely on this principle; so do regulators, rectifiers, mechanical governors, and electromagnetic emitters. Electric motor/generators and heat engines are rotary variants of this process; they use cybernetic phase alternation to ensure that the crankshaft constantly tries to catch up to itself.

Paradox, in the form of negative-feedback loops, is at the heart of all high technology. Since mechanical Liars (and Fools) dominate modern life, let us investigate their logic.

B.   Diamond Values

Consider the period-2 oscillations of binary values.
There are four such logic waves:

t t t t t t .... ;   call this "t/t", or "t".

t f t f t f .... ;   call this "t/f", or "i".

f t f t f t .... ;   call this "f/t", or "j".

f f f f f f .... ;   call this "f/f", or "f".

"/" is pronounced "but"; thus i is "true but false" and j is "false but true". These four values form a diamond-shaped lattice:

$$\textbf{true = t/t}$$

$$\textbf{i = t/f} \qquad\qquad \textbf{j = f/t}$$

$$\textbf{false = f/f}$$

This is "diamond logic"; a wave logic with two components and four truth values. It describes the logic waves of period 2.

C. Harmonic Functions

Let the positive operators $\wedge$ ("and") and $\vee$ ("or") operate termwise:

$$(a/b) \wedge (c/d) = (a \wedge c)/(b \wedge d),$$

$$(a/b) \vee (c/d) = (a \vee c)/(b \vee d).$$

We can then define "but" as a projection operator:

$$a/b = (a \wedge i) \vee (b \wedge j)$$

$$= (a \vee j) \wedge (b \vee i).$$

In diamond logic, negation operates after a flip:

$$\sim(a/b) = (\sim b)/(\sim a).$$

This corresponds to a split-second time delay in evaluating negation; and this permits fixedpoints:

$$\sim(t/f) = (\sim f)/(\sim t) = t/f,$$

$$\sim(f/t) = (\sim t)/(\sim f) = f/t.$$

Thus paradox is possible in diamond logic.

Call a function "harmonic" if it can be defined from $\wedge$, $\vee$, $\sim$, and the four values t, i, j, f. They include:

$$(a \Rightarrow b) = (\sim a) \vee b,$$

$$(a \text{ iff } b) = (\sim a \vee b) \wedge (\sim b \vee a),$$

$$(a \text{ xor } b) = (a \wedge \sim b) \vee (b \wedge \sim a).$$

The "majority" operator M has two definitions:

$$M(a, b, c) = (a \wedge b) \vee (b \wedge c) \vee (c \wedge a)$$

$$= (a \vee b) \wedge (b \vee c) \wedge (c \vee a).$$

Here are the "lattice operators":

$$a \min b = (a \vee b)/(a \wedge b) = \text{``}a \vee / \wedge b\text{''}.$$

$$a \max b = (a \wedge b)/(a \vee b) = \text{``}a \wedge / \vee b\text{''}.$$

We can define "but" from the lattice operators:

$$a/b = (a \min \mathrm{f}) \max(b \min \mathrm{t}) = (a \max \mathrm{t}) \min(b \max \mathrm{f}).$$

Here are the two "harmonic projection" operators:

$$\lambda(x) = x/\sim x,$$

$$\rho(x) = \sim x/x.$$

Here are the upper and lower differentials:

$$\mathrm{D}x = (x \Rightarrow x) = (x \text{ iff } x) = (x \vee \sim x),$$

$$\mathrm{d}x = (x - x) = (x \text{ xor } x) = (x \wedge \sim x).$$

This, then, is Diamond; a logic containing the boolean values, plus paradoxes and lattice operators.

Here are truth tables for the functions defined above:

x:	~x:	∧ y: t f i j	∨ y: t f i j	⇒ y: t f i j	iff y: t f i j	xor y: t f i j
t	f	t f i j	t t t t	t f i j	t f i j	f t i j
f	t	f f f f	t f i j	t t t t	f t i j	t f i j
i	i	i f i f	t i i t	t i i t	i i i t	i i i f
j	j	j f f j	t j t j	t j t j	j j t j	j j f j

x:	but y: t f i j	min y: t f i j	max y: t f i j	λx:	ρx:	Dx:	dx:	M(x, y, z) majority
t	t i i t	t i i t	t j t j	i	j	t	f	y ∨ z
f	j f f j	i f i f	j f f j	j	i	t	f	y ∧ z
i	t i i t	i i i i	t f i j	i	i	i	i	y min z
j	j f f j	t f i j	j j j j	j	j	j	j	y max z

D.   Gluts and Gaps

The values i and j can be interpreted as "underdetermined" and "overdetermined"; where "underdetermined" means "insufficient data for definite answer", and "overdetermined" means "contradictory data". Thus, an underdetermined statement is neither provable nor refutable, and an overdetermined statement is both provable and refutable.

Underdetermined can also be called "gap", i.e. neither true nor false; and overdetermined can be called "glut", i.e. both true and false. We therefore say:

True is true = true;	True is false = false;
False is true = false;	False is false = true;
Gap is true = false;	Gap is false = false;
Glut is true = true;	Glut is false = true;

Let us assume that values are equal if they are equally true and equally false. That is:

If

$(A$ is true$) = (B$ is true$)$ and $(A$ is false$) = (B$ is false$)$

then

$A = B.$

We can then define the logical operators thus:

$$(A \wedge B) \text{ is true } = (A \text{ is true}) \wedge (B \text{ is true}),$$
$$(A \wedge B) \text{ is false} = (A \text{ is false}) \vee (B \text{ is false}),$$
$$(A \vee B) \text{ is true } = (A \text{ is true}) \vee (B \text{ is true}),$$
$$(A \vee B) \text{ is false} = (A \text{ is false}) \wedge (B \text{ is false}),$$
$$(\sim A) \text{ is true } \quad = (A \text{ is false}),$$
$$(\sim A) \text{ is false } \quad = (A \text{ is true}) \,.$$

It immediately follows from these definitions that:

$$A \wedge B \quad = B \wedge A,$$
$$A \vee B \quad = B \vee A,$$
$$A \wedge F \quad = F,$$
$$A \vee T \quad = T,$$
$$A \wedge T \quad = A \vee F = A,$$
$$A \wedge A \quad = A \vee A = A,$$
$$\text{glut} \wedge \text{gap} = F,$$
$$\text{glut} \vee \text{gap} = T,$$
$$\sim T \quad = F,$$
$$\sim F \quad = T,$$
$$\sim \text{glut} \ = \text{glut},$$
$$\sim \text{gap} \ = \text{gap}.$$

These equations imply this table:

x:	~x:		∧y:					∨y:			
			t	f	gp	gl		t	f	gp	gl
t	f		t	f	gp	gl		t	t	t	t
f	t		f	f	f	f		t	f	gp	gl
gp	gp		gp	f	gp	f		t	gp	gp	t
gl	gl		gl	f	f	gl		t	gl	t	gl

This is equivalent to the diamond tables under two interpretations:

True = t; False = f; Gap = i; Glut = j;
True = t; False = f; Gap = j; Glut = i.

According to Gödel's Theorem, any logic system is either incomplete or inconsistent; thus the equation;

$$i \lor j = t$$

that is; underdetermined or overdetermined = true — is none other than Gödel's Theorem, written as a diamond equation. Diamond harmonizes with meta-mathematics.

I and J are complementary paradoxes; the yin and yang of diamond logic. They oppose, yet reflect.

Yang is not yang, yin is not yin, and the Tao is not the Tao!

E. Diamond Circuits

One can implement diamond logic in switching circuits, two different ways; via "phased delay" and via "dual rail".

In "phased delay", one lets a switching circuit oscillate. T then means "on", F means "off", and the two mid-values mean the two wobbles of opposite phase. This implementation requires a global clock to ensure that the switches wobble in synchrony; it also reqires that all inverter switches take a unit delay, and all positive gates ($\wedge$, $\vee$, majority) take delay two:

$$(\sim B)(n) = \sim(B(n-1)),$$

$$(A \wedge B)(n) = (A(n-2) \wedge B(n-2)),$$

$$(A \vee B)(n) = (A(n-2) \vee B(n-2)),$$

$$(M(A, B, C))(n) = (M(A(n-2), B(n-2), C(n-2))).$$

In a "dual rail" circuit, all wires in a standard switching circuit are replaced by pairs of wires. True then means "both rails on", false means "both rails off", and the two mid-values mean that one of the two rails is on.

The gates then are:

"not" :

"and":

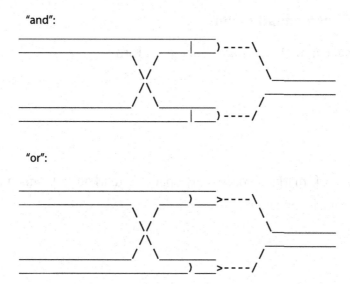

"or":

Later we will see a "Ganglion", which uses a "phased rail" system.

F.   Brownian Forms

Make a mark. This act generates a form:

A mark marks a space. Any space, if marked, remains marked if the mark is repeated:

where "=" denotes "is confused with".

This is the crossed form, or "mark".

Each mark is a call; to recall is to call.

A mark is a crossing, between marked and unmarked space. To cross twice is not to cross; thus a mark within a mark is indistinguishable from an unmarked space:

This is the uncrossed form, or "void".

Each mark is a crossing; to recross is not to cross.

Thus we get the "arithmetic initials" for George Spencer-Brown's famous Laws of Form. In his remarkable book, *Laws of Form*, Spencer-Brown demonstrated that these suffice to evaluate all formal expressions in the Brownian calculus; and that these forms obey two "algebraic initials":

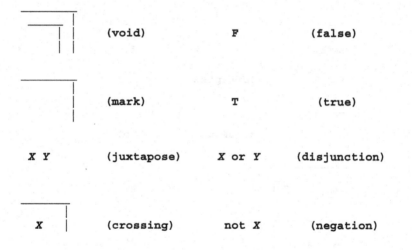

He proved that these axioms are consistent, independent and complete; that is, they prove all arithmetic identities. This "primary algebra" can be identified with Boolean logic. The usual matching is:

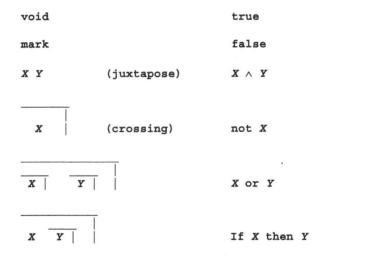

$X \wedge Y$ (conjunction)

If X, then Y

X xor Y

Majority (X, Y, Z)

There is a complementary interpretation:

void	true
mark	false
$X\ Y$ (juxtapose)	$X \wedge Y$
X (crossing)	not X
X Y (or)	X or Y
X Y	If X then Y

The standard interpretation is usually preferred because it has a simpler implication operator.

We can extend the Brownian calculus to diamond by introducing two new forms: "curl" and "uncurl":

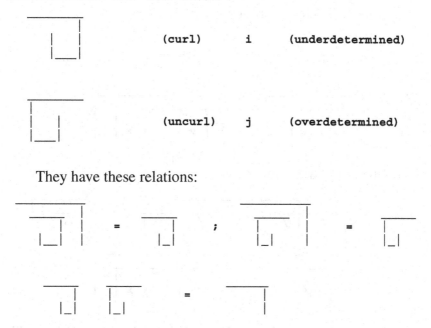

They have these relations:

You can create other interpretations simply by swapping the roles of curl and uncurl. (This exploits the symmetries of the diamond. See the "Conjugation" chapter below.)

In this interpretation, we have:

$$\overline{\overline{x} \mid x \mid} \quad = \quad dx \quad ; \quad \overline{x} \mid \overline{x} \quad = \quad Dx \qquad : \text{Differentials}$$

$(x \min y) = M(x, \text{curl}, y)$

$$= \quad \overline{\overline{x \mid \backslash\_\mid} \mid} \quad \overline{\overline{\backslash\_\mid y} \mid} \quad \overline{y \mid x \mid}$$

$$= \quad \overline{x \; \overline{\backslash\_\mid} \mid} \quad \overline{\overline{\backslash\_\mid y}} \quad \overline{y \, x \mid}$$

$(x \max y) = M(x, \text{uncurl}, y)$

$$= \quad \overline{\overline{x \mid \mid\_/} \mid} \quad \overline{\overline{\mid\_/ \; y} \mid} \quad \overline{y \mid x \mid}$$

$$= \quad \overline{x \; \overline{\mid\_/} \mid} \quad \overline{\overline{\mid\_/ \; y}} \quad \overline{y \, x \mid}$$

When one includes the new Brownian forms "curl" and "uncurl", these axioms still hold:

$$\overline{\overline{A} \mid \overline{B} \mid} C \quad = \quad \overline{\overline{A\,C} \mid \overline{B\,C} \mid} \quad ; \quad \textbf{"Transposition"}$$

$$\overline{A \mid \overline{B} \mid} A \quad = \quad A \quad ; \quad \textbf{"Occultation"}$$

We also get this equation; "**Complementarity**":

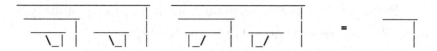

— a two-component anti-boolean axiom. These three axioms suffice to calculate all the truth tables, and all the algebraic identities, of diamond.

G. Boundary Logic

Boundary logic is Brownian form algebra, adapted for the typewriter. It uses brackets instead of Brown's mark:

[A] instead of $\overline{A|}$.

The arithmetic initials are then:

[ ][ ] = [ ].

[[ ]] = 1.

If we call [] "1" and [[]] "0", then we get these equations:

[0] = 1;   [1] = 0;

  0 0 = 0;   0 1 = 1 0 = 1 1 = 1.

The algebraic initials are:

[[a][b]]c = [[ac][bc]].

  [[a] a] = .

We can identify boundary forms with boolean logic this way:

[ ]	=	true;
[[ ]]	=	false;
[A]	=	$\sim A$;
$A\ B$	=	$A \vee B$;
[[A] [B]]	=	$A \wedge B$;
[A] B	=	if A then B;
[[A] B] [[B] A]	=	A xor B;

$$[[[A] B] [[B] A]] \quad = \quad A \text{ iff } B;$$
$$[[A B] [B C] [C A]] \quad = \quad \text{Majority } (A, B, C)$$

For diamond, we introduce two new expressions; 6 and 9:

$$[6] = 6; [9] = 9; 69 = [\,].$$

We can identify 6 with i, and 9 with j. (Or vice versa.)

We then get these equations;

$$10 = 16 = 19 = 11 = 1; [1] = 0;$$

$$91 = 96 = 1; 99 = 90 = 9; [9] = 9;$$

$$61 = 69 = 1; 66 = 60 = 6; [6] = 6;$$

$$00 = 0; 06 = 6; 09 = 9; 01 = 1; [0] = 1.$$

Inspection of tables shows that juxtaposition **ab** is isomorphic to diamond disjunction; and crossing **[a]** is isomorphic to diamond negation.

For the four forms 0, 1, 6, and 9, we get these identities:

Transposition:	$[[a] [b]] c$	$=$	$[[a c] [b c]]$.
Occultation:	$[[a] b] a$	$=$	a.
Complementarity:	$[[6] 6] [[9] 9]$	$=$	1.

In the next chapter, we'll see that these axioms define diamond logic.

Chapter 3

Diamond Algebra

Bracket Algebra
Laws
Normal Forms
Completeness

A. Bracket Algebra

Call these the bracket axioms:

Transposition: $[[a][b]]c = [[a\ c][b\ c]]$.

Occultation: $[[a]b]a = a$.

Complementarity: $[[6]6][[9]9] = [\ ]$.

In addition, we assume commutativity and associativity for juxtaposition:

$$a\ b = b\ a; \quad a\ b\ c = a\ b\ c.$$

These equations are implicit in the bracket notation. Brackets distinguish only inside from outside, not left from right.

From the bracket axioms we can derive theorems:

Reflexion. $[[x]] = x$.

Proof.

$$[[x]] = [[x][[\ ]x]] \quad \text{occ.}$$
$$= [[\ ][[\ ]]]x \quad \text{trans.}$$
$$= x \quad \text{occ.}$$

Identity. $[[\;]]x = x$.

Proof. Directly from Occultation.

Domination. $[\;]x = [\;]$.

Proof. $[\;]x = [[[\;]x]] = [\;]$ ref., occ.

Recall. $xx = x$.

Proof. $xx = [[x]]x = x$ ref., occ.

Duality. $69 = [6]9 = 6[9] = [6][9] = [\;]$

Proof.

$$[\;] = [[6]6] \, [[9]9] \qquad\qquad\qquad\qquad\text{comp.}$$
$$[\;] = [[6] \, [[6]]] \, [[9] \, [[9]]] \qquad\qquad\qquad\text{ref., twice}$$
$$[\;] = [[6[[9] \, [[9]]]] \, [[6][[9] \, [[9]]]]] \qquad\text{trans.}$$
$$[\;] = [[[[69][6[9]]]] \, [[[6][9] \, [[6][9]]]]] \qquad\text{trans.}$$
$$[\;] = [[69] \, [6[9]] \, [6] \, [9] \, [[6][9]]] \qquad\text{ref., twice}$$
$$[[\;]] = [69] \, [6[9]] \, [6] \, [9] \, [[6][9]] \qquad\qquad\text{ref.}$$

$$\text{Ergo } [69] = [[\;]] \, [69]$$
$$= [6[9]] \, [6] \, [9] \, [[6][9]] \, [69] \, [69]$$
$$= [6[9]] \, [6] \, [9][[6] \, [9]] \, [69] = [[\;]]$$

So $[69] = [[\;]]$; similarly $[[6]9] = [6[9]] = [[6][9]] = [[\;]]$;
Therefore $69 = [6]9 = 6[9] = [6][9] = [\;]$.

Fixity. $[6] = 6$; $[9] = 9$.

Proof.

$$6 = 6[[\;]] = 6[6[9]] = 6[[[6]] [9]] = [[6[6]] [69]]$$
$$= [[6[6]] [[\;]]] = [[6[6]][[6]9]] = [[6] [9]] [6] = [6].$$

Similarly, $9 = [9]$.

From Fixity, Reflexion, Domination, Duality, Recall, and Identity, we can derive the tables for $[a]$ and ab. The bracket axioms yield
the bracket arithmetic.

Reoccultation. $[xy][x] = [x]$.

Proof.

$$[xy][x] = [[[x]]y][x] \quad \text{ref.}$$
$$= [x] \qquad\qquad \text{occ.}$$

Echelon. $[[[x]y]z] = [xz][[y]z]$.

Proof.

$$[[[x]y]z] = [[[x][[y]]]z] \quad \text{ref.}$$
$$= [[[xz][[y]z]]] \quad \text{trans.}$$
$$= [xz][[y]z] \quad \text{ref.}$$

Modified Generation. $[[xy]y] = [[x]y][y[y]]$.

Proof.

$$[[xy]y] = [[[[x]][[y]]]y] \quad \text{ref.}$$
$$= [[[[x]y][[y]y]]] \quad \text{trans.}$$
$$= [[x]y][[y]y] \quad \text{ref.}$$

Modified Extension. $[[x]y][[x][y]] = [[x][y[y]]]$.

Proof.

$$\begin{aligned}
[[x]y][[x][y]] &= [[[[x]y][[x][y]]]] &&\text{ref.}\\
&= [[[y][[y]]][x]] &&\text{trans.}\\
&= [[x][y[y]]] &&\text{ref.}
\end{aligned}$$

Inverse Transposition. $[[xy][z]] = [[x][z]][[y][z]]$.

Proof.

$$\begin{aligned}
[[xy][z]] &= [[[[x]][[y]]][z]] &&\text{ref.}\\
&= [[[[x][z]][[y][z]]]] &&\text{trans.}\\
&= [[x][z]][[y][z]] &&\text{ref.}
\end{aligned}$$

Modified Transposition. $[[x][yw][zw]] = [[x][y][z]][[x][w]]$.

Proof.

$$\begin{aligned}
[[x][yw][zw]] &= [[x][[[yw][zw]]]] &&\text{ref.}\\
&= [[x][[[y][z]]w]] &&\text{trans.}\\
&= [[x][[[y][z]][[w]]]] &&\text{ref.}\\
&= [[[[x][y][z]][[x][w]]]] &&\text{trans.}\\
&= [[x][y][z]][[x][w]] &&\text{ref.}
\end{aligned}$$

Majority. $[[xy][yz][zx]] = [[x][y]][[y][z]][[z][x]]$.

Proof.

$$\begin{aligned}
[[xy][yz][zx]] &= [[xy][x][y]][[xy][z]] &&\text{mod.trans.}\\
&= [[x][y]][[xy][z]] &&\text{reocc.}\\
&= [[x][y]][[x][z]][[y][x]] &&\text{inv.trans.}
\end{aligned}$$

Retransposition (3 terms).

$$[a_1x][a_2x][a_3x] = [[[a_1][a_2][a_3]]x].$$

Proof.

$$
\begin{aligned}
[a_1x][a_2x][a_3x] &= [[[a_1x][a_2x]]][a_3x] && \text{ref.} \\
&= [[[a_1][a_2]]x][a_3x] && \text{trans.} \\
&= [[[[[a_1][a_2]]x][a_3x]]] && \text{ref.} \\
&= [[[[[a_1][a_2]]]][a_3]]x] && \text{trans.} \\
&= [[[a_1][a_2][a_3]]x] && \text{ref.}
\end{aligned}
$$

Retransposition (n terms).

$$[a_1x][a_2x]\cdots[a_nx] = [[[a_1][a_2]\cdots[a_n]]x].$$

Proof is by induction on n. Given that the theorem is true for n, the following proves it for $n + 1$:

$$
\begin{aligned}
[a_1x][a_2x]&\cdots[a_nx][a_{n+1}x] \\
&= [[[a_1][a_2]\cdots[a_n]]x][a_{n+1}x] && * \\
&= [[[[[a_1][a_2]\cdots[a_n]]x][a_{n+1}x]]] && \text{ref.} \\
&= [[[[[a_1][a_2]\cdots[a_n]]]][a_{n+1}]]x] && \text{trans.} \\
&= [[[a_1][a_2]\cdots[a_n][a_{n+1}]]x] && \text{ref.}
\end{aligned}
$$

Cross-Transposition.

$$[[[a]x][[b][x]][x[x]]] = [ax][b[x]][x[x]].$$

Proof.

$$
\begin{aligned}
[[[a]x]&[[b][x]][x[x]]] \\
&= [[[a]x][[b][x]][[[x]][x]]] && \text{ref.} \\
&= [[[[[a]x][[b][x]][x]][[[a]x][[b][x]]x]]] && \text{trans.} \\
&= [[[a]x][[b][x]][x]][[[a]x][[b][x]]x] && \text{ref.}
\end{aligned}
$$

$$= [[[a][[x]]][[b][x]][x]] \, [[[a]x][[b][x]]x] \quad \text{ref.}$$

$$= [[[b][x]][x]] \, [[[a]x]x] \quad \text{occ.}$$

$$= [[[b][x]][x]] \, [[[a][[x]]]x] \quad \text{ref.}$$

$$= [[[b[x]][x[x]]]] \, [[[ax][[x]x]]] \quad \text{trans.}$$

$$= [b[x]][x[x]] \, [ax][[x]x] \quad \text{ref.}$$

$$= [ax][b[x]] \, [x[x]] \quad \text{recall.}$$

This result translates into diamond logic in two dual ways:

$$(A \wedge x) \vee (B \wedge \sim x) \vee \mathrm{d}x = (A \vee \sim x) \wedge (B \vee x) \wedge \mathrm{D}x,$$

$$(a \vee x) \wedge (b \vee \sim x) \wedge \mathrm{D}x = (a \wedge \sim x) \vee (b \wedge x) \vee \mathrm{d}x.$$

General Cross-Transposition.

$$[[a][x]] \, [[b]x] \, [[c][x]x] = [[a[x]][bx][abc][x[x]]].$$

Proof. From right to left.

$[[a[x]][bx][abc][x[x]]]$

$$= [[[[a[x]][bx][x[x]]]][abc]] \quad \text{ref.}$$

$$= [[[[a][x]][[b]x][x[x]]]][abc]] \quad \text{crosstrans.}$$

$$= [[[[a][x][abc]][[b]x[abc]][x[x][abc]]]] \quad \text{trans.}$$

$$= [[a][x][abc]][[b]x[abc]][x[x][abc]] \quad \text{ref.}$$

$$= [[a][x]][[b]x][x[x][abc]] \quad \text{reocc.}$$

$$= [[a][x]][[b]x][x[x][[[a]][[b]][[c]]]] \quad \text{ref.}$$

$$= [[a][x]][[b]x][[[x[x][a]][x[x][b]][x[x][c]]]] \quad \text{trans.}$$

$$= [[a][x]][[b]x][x[x][a]][x[x][b]][x[x][c]] \quad \text{ref.}$$

$$= [[a][x]][[b]x][x[x][c]] \quad \text{reocc.}$$

This result translates into diamond logic in two ways:

$$(A \wedge x) \vee (B \wedge \sim x) \vee (C \wedge \mathrm{d}x)$$

$$= (A \vee \sim x) \wedge (B \vee x) \wedge (A \vee B \vee C) \wedge \mathrm{D}x,$$

$$(a \vee x) \wedge (b \vee \sim x) \wedge (c \vee \mathrm{D}x)$$

$$= (a \wedge \sim x) \vee (b \wedge x) \vee (a \wedge b \wedge c) \vee \mathrm{d}x.$$

Here are two examples of general cross-transposition:

$$a \text{ xor } b = (a \wedge \sim b) \vee (\sim a \wedge b) \vee (b \wedge \sim b \wedge f)$$

$$= (a \vee b) \wedge (\sim a \vee \sim b) \wedge (a \vee \sim a \vee f) \wedge \mathrm{D}b$$

$$= (a \vee b) \wedge (\sim a \vee \sim b) \wedge \mathrm{D}a \wedge \mathrm{D}b$$

$$= (a \text{ iff } \sim b) \wedge \mathrm{D}a \wedge \mathrm{D}b,$$

$$a \text{ iff } b = (a \vee \sim b) \wedge (\sim a \vee b) \wedge (b \vee \sim b \vee t)$$

$$= (a \wedge b) \vee (\sim a \wedge \sim b) \vee (a \wedge \sim a \wedge t) \vee \mathrm{d}b$$

$$= (a \wedge b) \vee (\sim a \wedge \sim b) \vee \mathrm{d}a \vee \mathrm{d}b$$

$$= (a \text{ xor } \sim b) \vee \mathrm{d}a \vee \mathrm{d}b.$$

Let $M(x, y, z)$ denote $[[xy][yz][zx]]$, or $[[x][y]][[y][z]][[z][x]]$. We can derive these theorems:

Transmission. $[M(x, y, z)] = M([x], [y], [z])$.

Proof.

$$
\begin{aligned}
[M(x, y, z)] &= [[[xy][yz][zx]]] & \text{def.} \\
&= [xy][yz][zx] & \text{ref.} \\
&= [[[x]][[y]]]\,[[[y]][[z]]]\,[[[z]][[x]]] & \text{ref.} \\
&= M([x], [y], [z]) & \text{def.}
\end{aligned}
$$

Distribution. $x\mathrm{M}(a, b, c) = \mathrm{M}(xa, b, xc)$.

Proof.

$$
\begin{aligned}
x\mathrm{M}(a, b, c) &= x[[ab][bc][ca]] &&\text{def.} \\
&= [[xab][xbc][xca]] &&\text{trans.} \\
&= [[xab][bxc][xcxa]] &&\text{recall.} \\
&= \mathrm{M}(xa, b, xc) &&\text{def.}
\end{aligned}
$$

Redistribution. $[[x][\mathrm{M}(a, b, c)]] = \mathrm{M}([[x][a]], b, [[x][c]])$.

Proof.

$$
\begin{aligned}
[[x][\mathrm{M}(a, b, c)]] &= [[x]\mathrm{M}([a], [b], [c])] &&\text{trans.} \\
&= [\mathrm{M}([x][a], [b], [x][c])] &&\text{dist.} \\
&= \mathrm{M}([[x][a]], [[b]], [[x][c]]) &&\text{trans.} \\
&= \mathrm{M}([[x][a]], b, [[x][c]]) &&\text{ref.}
\end{aligned}
$$

Collection. $\mathrm{M}(x, y, z)) = [[x][y]][[xy][z]]$.

Proof.

$$
\begin{aligned}
\mathrm{M}(x, y, z)) &= [[xy][xz][yz]] &&\text{def.} \\
&= [[xy][x][y]][[xy][z]] &&\text{mod.trans.} \\
&= [[x][y]][[xy][z]] &&\text{reocc.}
\end{aligned}
$$

General Distribution. $\mathrm{M}(x, y, \mathrm{M}(a, b, c))$
$$= \mathrm{M}(\mathrm{M}(x, y, a), b, \mathrm{M}(x, y, c)).$$

Proof.

$$
\begin{aligned}
\mathrm{M}(x, y, \mathrm{M}(a, b, c)) & \\
&= [[x][y]][[xy][\mathrm{M}(a, b, c)]] &&\text{collect.} \\
&= [[x][y]]\mathrm{M}([[xy][a]], b, [[xy][c]]) &&\text{redist.} \\
&= \mathrm{M}([[x][y]][[xy][a]], b, [[x][y]][[xy][c]]) &&\text{dist.} \\
&= \mathrm{M}(\mathrm{M}(x, y, a), b, \mathrm{M}(x, y, c)) &&\text{collect.}
\end{aligned}
$$

Coalition. $M(x, x, y) = x$.

Proof.

$$M(x, x, y) = [[xx][xy][yx]] \quad \text{def.}$$
$$= [[x][xy]] \quad\quad \text{recall.}$$
$$= [[x]] \quad\quad\quad \text{reocc.}$$
$$= x \quad\quad\quad\quad \text{ref.}$$

General Associativity. $M(x, a, M(y, a, z)) = M(M(x, a, y), a, z)$.

Proof.

$$M(x, a, M(y, a, z)) = M(M(x, a, y), M(x, a, a), z) \quad \text{g.dist.}$$
$$= M(M(x, a, y), a, z) \quad\quad\quad \text{coal.}$$

These results prove that these operators:

$$M(x, [[\ ]], y) = [[x][y]];$$

$$M(x, [\ ], y) = xy;$$

$$M(x, 6, y) = [[x6][y6][xy]];$$

$$M(x, 9, y) = [[x9][y9][xy]]$$

have these properties: associativity; recall; attractors ($[[\ ]]$, $[\ ]$, 6 and 9, respectively); and mutual distribution.

B. Laws

Diamond obeys these *De Morgan* laws:

Commutativity: $A \vee B = B \vee A; A \wedge B = B \wedge A.$

Associativity: $(A \wedge B) \wedge C = A \wedge (B \wedge C),$

$(A \vee B) \vee C = A \vee (B \vee C).$

Distributivity: $A \wedge (B \vee C) = (A \wedge B) \vee (A \wedge C),$

$A \vee (B \wedge C) = (A \vee B) \wedge (A \vee C).$

Identities: $A \wedge t = A; A \vee f = A.$

Attractors: $A \wedge f = f; A \vee t = t.$

Recall: $A \wedge A = A; A \vee A = A.$

Absorption: $A \wedge (A \vee B) = A; A \vee (A \wedge B) = A.$

Double Negation: $\sim(\sim A) = A.$

De Morgan: $\sim(A \wedge B) = (\sim A) \vee (\sim B),$

$\sim(A \vee B) = (\sim A) \wedge (\sim B).$

These are the boolean laws, minus the Law of the Excluded Middle. In addition to these laws, Diamond has this "anti-boolean" law:

Complementarity: $Di \wedge Dj = f; \quad di \vee dj = t.$

That is:

$(i \vee \sim i) \wedge (j \vee \sim j) = f.$

$(i \wedge \sim i) \vee (j \wedge \sim j) = t.$

The Diamond laws suffice to prove:

Duality:

$$i \wedge j = (\sim i) \wedge j = (\sim j) \wedge i = (\sim i) \wedge (\sim j) = f,$$

$$i \vee j = (\sim i) \vee j = (\sim j) \vee i = (\sim i) \vee (\sim j) = t.$$

Fixity:

$$\sim i = i; \quad \sim j = j.$$

These rules, plus Identity, Attractors and Recall, suffice to construct diamond's truth tables.

The dual paradoxes i and j define "but", the "junction" operator:

$$x/y = (x \wedge i) \vee (y \wedge j) = (x \vee j) \wedge (y \vee i).$$

Here are the **junction** laws:

Recall:	a/a	$= a.$
Polarity:	$(a/b)/(c/d)$	$= a/d.$
Parallellism:	$(a/b) \wedge (c/d)$	$= (a \wedge c)/(b \wedge d),$
	$(a/b) \vee (c/d)$	$= (a \vee c)/(b \vee d),$
	$M(a/A, b/B, c/C)$	$= M(a, b, c)/M(A, B, C).$
Reflection:	$\sim(a/b)$	$= (\sim b)/(\sim a).$

Min and max obey these **lattice** laws:

Commutativity: $x \min y = y \min x$; $x \max y = y \max x$.

Associativity: $x \min (y \min z) = (x \min y) \min z$,

$x \max (y \max z) = (x \max y) \max z$.

Absorption: $x \max (x \min y) = x \min (x \max y) = x$.

Recall: $x \min x = x \max x = x$.

Attractors: $x \min i = i$; $x \max j = j$.

Identities: $x \min j = x$; $x \max i = x$.

Transmission: $\sim(x \min y) = (\sim x) \min (\sim y)$,

$\sim(x \max y) = (\sim x) \max (\sim y)$.

Mutual Distribution: $x^{**}(y{+}{+}z) = (x^{**}y){+}{+}(x^{**}z)$,

where $**$ and $++$ are both from: $\{\wedge, \vee,$

$\min, \max\}$.

The operators $\lambda(x)$ and $\rho(x)$ have these **harmonic projection** laws:

Duality: $\lambda(\sim x) = \rho(x)$; $\rho(\sim x) = \lambda(x)$.

Fixity: $\sim\lambda(x) = \lambda(x)$; $\sim\rho(x) = \rho(x)$.

Transmission: $\lambda(M(x, y, z)) = M(\lambda(x), \lambda(y), \lambda(z))$,

$\rho(M(x, y, z)) = M(\rho(x), \rho(y), \rho(z))$,

$\lambda(x \wedge y) = \lambda(x \max y) = \lambda(x) \max \lambda(y)$,

$\lambda(x \vee y) = \lambda(x \min y) = \lambda(x) \min \lambda(y)$,

$\rho(x \wedge y) = \rho(x \min y) = \rho(x) \min \rho(y)$,

$\rho(x \vee y) = \rho(x \max y) = \rho(x) \max \rho(y)$.

Bias: $\lambda(x/y) = \lambda(x)$; $\rho(x/y) = \rho(y)$.

Differentials: $\lambda(x) \vee \rho(x) = Dx$; $\lambda(x) \wedge \rho(x) = dx$.

Here are some **majority** laws:

Modulation: $M(a, f, b) = a \wedge b,$

$M(a, t, b) = a \vee b,$

$M(a, i, b) = a \min b,$

$M(a, j, b) = a \max b.$

Symmetry: $M(x, y, z) = M(y, z, x) = M(z, x, y) = M(x, z, y)$

$= M(z, y, x) = M(y, x, z).$

Coalition: $M(x, x, y) = M(x, x, x) = x.$

Transmission: $\sim(M(x, y, z)) = M(\sim x, \sim y, \sim z).$

Distribution: $M(a, b, M(c, d, e)) = M(M(a, b, c), d, M(a, b, e)).$

Modulation plus Transmission explains DeMorgan and Transmission:

$$\sim(a \wedge b) = (\sim a) \vee (\sim b),$$

$$\sim(a \vee b) = (\sim a) \wedge (\sim b),$$

$$\sim(a \min b) = (\sim a) \min(\sim b),$$

$$\sim(a \max b) = (\sim a) \max(\sim b).$$

The differentials dx and Dx obey these **derivative** laws:

$dx = dx \wedge x = dx \wedge Dx,$

$x = x \vee dx = x \wedge Dx,$

$Dx = Dx \vee x = Dx \vee dx,$

i.e. dx is a subset of x, which is a subset of Dx.

In Venn diagram terms, dx is the boundary of x, and Dx is everything else.

I call the following the **Leibnitz rules**, due to their similarity to the Leibnitz rule for derivatives of products:

$$d(x \wedge y) = (dx \wedge y) \vee (x \wedge dy),$$
$$D(x \vee y) = (Dx \vee y) \wedge (x \vee Dy),$$
$$d(x \vee y) = (dx \wedge \sim y) \vee (\sim x \wedge dy),$$
$$D(x \wedge y) = (Dx \vee \sim y) \wedge (\sim x \vee Dy).$$

Here are more **differential logic** identities:

$$ddx = dDx = dx,$$
$$DDx = Ddx = Dx,$$
$$\sim dx = Dx; \sim Dx = dx,$$
$$d(\sim x) = dx; D(\sim x) = Dx,$$
$$dx = (x - x) = x \text{ xor } x,$$
$$Dx = (x \Rightarrow x) = x \text{ iff } x,$$

$$d(x - y) = D(x \Rightarrow y) = (\sim y \wedge dx) \vee (x \wedge dy),$$
$$D(x - y) = d(x \Rightarrow y) = (y \vee Dx) \wedge (\sim x \vee Dy),$$
$$d(x \text{ xor } y) = d(x \text{ iff } y) = dx \text{ xor } dy = Dx \text{ xor } Dy,$$
$$D(x \text{ xor } y) = D(x \text{ iff } y) = Dx \text{ iff } Dy = dx \text{ iff } dy,$$

$$dM(x, y, z) = ((y \text{ xor } z) \wedge dx) \vee ((z \text{ xor } x) \wedge dy)$$
$$\vee ((x \text{ xor } y) \wedge dz) \vee M(dx, dy, dz),$$
$$DM(x, y, z) = ((y \text{ iff } z) \vee Dx) \wedge ((z \text{ iff } x) \vee Dy)$$
$$\wedge ((x \text{ iff } y) \vee Dz) \wedge M(Dx, Dy, Dz),$$

$$d(x \min y) = dx \min dy \min((x \text{ xor } y) \max t),$$

$$d(x \max y) = dx \max dy \max((x \text{ xor } y) \min t),$$

$$D(x \min y) = dx \min dy \min((x \text{ iff } y) \max f),$$

$$D(x \max y) = dx \max dy \max((x \text{ iff } y) \min f),$$

$$d(x/y) = (x - y)/(y - x) = \lambda(x) \wedge \rho(y),$$
$$D(x/y) = (y \Rightarrow x)/(x \Rightarrow y) = \lambda(x) \vee \rho(y),$$

$$dx = (x/y - y/x)/(y/x - x/y),$$
$$Dx = (y/x \Rightarrow x/y)/(x/y \Rightarrow y/x),$$

$$x - y = d(x/y)/d(y/x),$$
$$x \Rightarrow y = D(y/x)/D(x/y).$$

Xor and iff have these **diagonal operator** laws:

$$a \text{ xor } b = \sim(a \text{ iff } b),$$

$$a \text{ xor } b = b \text{ xor } a,$$

$$a \text{ xor } b = d(a/b) \vee d(b/a),$$

$$a \text{ xor } b = (a \text{ iff } \sim b) \wedge Da \wedge Db,$$

$$a \text{ iff } b = b \text{ iff } a,$$

$$a \text{ iff } b = D(a/b) \wedge D(b/a),$$

$$a \text{ iff } b = (a \text{ xor } \sim b) \vee da \vee db,$$

$$(a \wedge c) \text{ xor } (b \wedge c) = ((a \text{ xor } b) \wedge c) \vee ((a \vee b) \wedge dc),$$

$$(a \vee c) \text{ iff } (b \vee c) = ((a \text{ iff } b) \vee c) \wedge ((a \wedge b) \vee Dc).$$

C. Normal Forms

By using the De Morgan laws, one can put any harmonic diamond function $F(x_1 \cdots x_n)$ into one of the following forms:

Disjunctive Normal Form:

$$F(x_1 \cdots x_n) = (t_{11}(x_1) \wedge t_{12}(x_2) \wedge \cdots \wedge t_{1n}(x_n))$$
$$\vee (t_{21}(x_1) \wedge t_{22}(x_2) \wedge \cdots \wedge t_{2n}(x_n))$$
$$\vee \cdots$$
$$\vee (t_{m1}(x_1) \wedge t_{m2}(x_2) \wedge \cdots \wedge t_{mn}(x_n)),$$

where each $t_{ij}(x_i)$ is one of these functions:

$$\{x_i, \sim x_i, \mathrm{d}x_i, \mathrm{t}, \mathrm{f}, \mathrm{i}, \mathrm{j}\}.$$

Conjunctive Normal Form:

$$F(x_1 \cdots x_n) = (t_{11}(x_1) \vee t_{12}(x_2) \vee \cdots \vee t_{1n}(x_n))$$
$$\wedge (t_{21}(x_1) \vee t_{22}(x_2) \vee \cdots \vee t_{2n}(x_n))$$
$$\wedge \cdots$$
$$\wedge (t_{m1}(x_1) \vee t_{m2}(x_2) \vee \cdots \vee t_{mn}(x_n)),$$

where each $t_{ij}(x_i)$ is one of these functions:

$$\{x_i, \sim x_i, \mathrm{D}x_i, \mathrm{t}, \mathrm{f}, \mathrm{i}, \mathrm{j}\}.$$

We do this by distributing negations downwards, canceling double-negations, and distributing enough times. These normal forms are just like their counterparts in boolean logic, except that they allow differential terms.

Theorem. The Primary Normal Forms

$$F(x) = (A \wedge x) \vee (B \wedge \sim x) \vee (C \wedge \mathrm{d}x) \vee \mathrm{D},$$
$$F(x) = (a \vee \sim x) \wedge (b \vee x) \wedge (c \vee \mathrm{D}x) \wedge d,$$

where A, B, C, D, a, b, c, d *are all free of variable* x, *and:*

$$A \vee D = F(\mathrm{t}) = a \wedge d,$$
$$B \vee D = F(\mathrm{f}) = b \wedge d,$$
$$A \vee B \vee C \vee D = F(\mathrm{i}) \vee F(\mathrm{j}) = d,$$
$$D = F(\mathrm{i}) \wedge F(\mathrm{j}) = a \wedge b \wedge c \wedge d.$$

Proof. We get the first two equations from the Disjunctive and Conjunctive Normal Forms by collecting like terms with respect to the variable x. The next two equations can be verified by substituting values t and f. Substituting i and j, plus using Paradox and Distribution, yields:

$$F(\mathrm{i}) = ((A \vee B \vee C) \wedge \mathrm{i}) \vee D = ((a \wedge b \wedge c) \vee \mathrm{i}) \wedge d,$$
$$F(\mathrm{j}) = ((A \vee B \vee C) \wedge \mathrm{j}) \vee D = ((a \wedge b \wedge c) \vee \mathrm{j}) \wedge d.$$

Therefore:

$$
\begin{aligned}
F(\mathrm{i}) \vee F(\mathrm{j}) &= ((A \vee B \vee C) \wedge \mathrm{i}) \vee D \vee ((A \vee B \vee C) \wedge \mathrm{j}) \vee D \\
&= ((A \vee B \vee C) \wedge (\mathrm{i} \vee \mathrm{j})) \vee D \\
&= A \vee B \vee C \vee D.
\end{aligned}
$$

Also:

$$
\begin{aligned}
F(\mathrm{i}) \wedge F(\mathrm{j}) &= (((A \vee B \vee C) \wedge \mathrm{i}) \vee D) \wedge (((A \vee B \vee C) \wedge \mathrm{j}) \vee D) \\
&= (((A \vee B \vee C) \wedge \mathrm{i}) \wedge (((A \vee B \vee C) \wedge \mathrm{j})) \vee D \\
&= ((A \vee B \vee C) \wedge \mathrm{i} \wedge \mathrm{j}) \vee D = D.
\end{aligned}
$$

Similarly:

$$F(\mathrm{i}) \wedge F(\mathrm{j}) = a \wedge b \wedge c \wedge d$$
$$F(\mathrm{i}) \vee F(\mathrm{j}) = d. \qquad\qquad\qquad \text{QED}$$

Now recall this theorem:

Cross-Transposition:

$$(A \wedge x) \vee (B \wedge \sim x) \vee (C \wedge \mathrm{d}x)$$
$$= (A \vee \sim x) \wedge (B \vee x) \wedge (A \vee B \vee C) \wedge \mathrm{D}x,$$
$$(a \vee x) \wedge (b \vee \sim x) \wedge (c \vee \mathrm{D}x)$$
$$= (a \wedge \sim x) \vee (b \wedge x) \vee (a \wedge b \wedge c) \vee \mathrm{d}x.$$

Second proof. Use Distribution, Occultation, and the derivative rules:

$$(A \wedge x) \vee (B \wedge \sim x) \vee (C \wedge \mathrm{d}x)$$
$$= (A \vee B \vee C) \wedge (A \vee B \vee \mathrm{d}x) \wedge (A \vee \sim x \vee C)$$
$$\wedge (A \vee \sim x \vee \mathrm{d}x) \wedge (x \vee B \vee C) \wedge (x \vee B \vee \mathrm{d}x)$$
$$\wedge (x \vee \sim x \vee C) \wedge (x \vee \sim x \vee \mathrm{d}x)$$
$$= (A \vee B \vee C) \wedge (A \vee B \vee \mathrm{d}x) \wedge (A \vee \sim x \vee C)$$
$$\wedge (A \vee \sim x) \wedge (x \vee B \vee C) \wedge (x \vee B)$$
$$\wedge (x \vee \sim x \vee C) \wedge (x \vee \sim x)$$
$$= (A \vee B \vee C) \wedge (A \vee B \vee \mathrm{d}x) \wedge (A \vee \sim x)$$
$$\wedge (x \vee B) \wedge (\mathrm{D}x \vee C) \wedge (\mathrm{D}x)$$
$$= (A \vee B \vee C) \wedge (A \vee B \vee x) \wedge (A \vee B \vee \sim x)$$
$$\wedge (A \vee \sim x) \wedge (B \vee x) \wedge (\mathrm{D}x)$$
$$= (A \vee B \vee C) \wedge (A \vee \sim x) \wedge (B \vee x) \wedge (\mathrm{D}x).$$

Similarly,

$$(a \vee x) \wedge (b \vee \sim x) \wedge (c \vee \mathrm{D}x)$$
$$= (a \wedge \sim x) \vee (b \wedge x) \vee (a \wedge b \wedge c) \vee \mathrm{d}x.$$

QED

The Primary Normal Forms, plus Cross-Transposition, yield:

Theorem. Differential Normal Forms

Any harmonic function $F(x)$ can be put into these forms:

$$F(x) = (F(t) \wedge x) \vee (F(f) \wedge \sim x) \vee M(F(i), dx, F(j)),$$

$$F(x) = (F(t) \vee \sim x) \wedge (F(f) \vee x) \wedge M(F(i), Dx, F(j)).$$

This separates the function into boolean and lattice components. It expresses the function in terms of its values.

Proof. Start with the Primary Normal Forms:

$$F(x) = (A \wedge x) \vee (B \wedge \sim x) \vee (C \wedge dx) \vee D$$

$$= (a \vee \sim x) \wedge (b \vee x) \wedge (c \vee Dx) \wedge d,$$

where A, B, C, D, a, b, c, d are all free of variable x, and

$$A \vee D = F(t) = a \wedge d,$$

$$B \vee D = F(f) = b \wedge d,$$

$$A \vee B \vee C \vee D = F(i) \vee F(j) = d,$$

$$D = F(i) \wedge F(j) = a \wedge b \wedge c \wedge d.$$

Then apply Cross-Transposition:

$$(A \wedge x) \vee (B \wedge \sim x) \vee (C \wedge dx)$$

$$= (A \vee \sim x) \wedge (B \vee x) \wedge (A \vee B \vee C) \wedge Dx,$$

$$(a \vee x) \wedge (b \vee \sim x) \wedge (c \vee Dx)$$

$$= (a \wedge \sim x) \vee (b \wedge x) \vee (a \wedge b \wedge c) \vee dx.$$

We derive:

$$F(x) = (A \wedge x) \vee (B \wedge \sim x) \vee (C \wedge \mathrm{d}x) \vee D$$
$$= ((A \vee \sim x) \wedge (B \vee x) \wedge (A \vee B \vee C) \wedge \mathrm{D}x) \vee D$$
$$= (A \vee D \vee \sim x) \wedge (B \vee D \vee x)$$
$$\wedge (A \vee B \vee C \vee \mathrm{D}) \wedge (\mathrm{D}x \vee D)$$
$$= (F(\mathrm{t}) \vee \sim x) \wedge (F(\mathrm{f}) \vee x)$$
$$\wedge (F(\mathrm{i}) \vee F(\mathrm{j})) \wedge (\mathrm{D}x \vee (F(\mathrm{i}) \wedge F(\mathrm{j})))$$
$$= (F(\mathrm{t}) \vee \sim x) \wedge (F(\mathrm{f}) \vee x)$$
$$\wedge (F(\mathrm{i}) \vee F(\mathrm{j})) \wedge (\mathrm{D}x \vee F(\mathrm{i})) \wedge (\mathrm{D}x \vee F(\mathrm{j}))$$
$$= (F(\mathrm{t}) \vee \sim x) \wedge (F(\mathrm{f}) \vee x) \wedge \mathrm{M}(F(\mathrm{i}), \mathrm{D}x, F(\mathrm{j})).$$

Dually:

$$F(x) = (a \vee x) \wedge (b \vee \sim x) \wedge (c \vee \mathrm{D}x) \wedge d$$
$$= ((a \wedge \sim x) \vee (b \wedge x) \vee (a \wedge b \wedge c) \vee \mathrm{d}x) \wedge d$$
$$= ((a \wedge d \wedge \sim x) \vee (b \wedge d \wedge x)$$
$$\vee (a \wedge b \wedge c \wedge d) \vee (\mathrm{d}x \wedge d)$$
$$= (F(\mathrm{f}) \wedge \sim x) \vee (F(\mathrm{t}) \wedge x)$$
$$\vee (F(\mathrm{i}) \wedge F(\mathrm{j})) \vee (\mathrm{d}x \wedge (F(\mathrm{i}) \vee F(\mathrm{j})))$$
$$= (F(\mathrm{t}) \wedge x) \vee (F(\mathrm{f}) \wedge \sim x)$$
$$\vee (F(\mathrm{i}) \wedge F(\mathrm{j})) \vee (\mathrm{d}x \wedge F(\mathrm{i})) \vee (\mathrm{d}x \wedge F(\mathrm{j}))$$
$$= (F(\mathrm{t}) \wedge x) \vee (F(\mathrm{f}) \wedge \sim x) \vee \mathrm{M}(F(\mathrm{i}), \mathrm{d}x, F(\mathrm{j})).$$

QED

D. Completeness

The Differential Normal Forms imply this theorem:

Completeness. Any equational identity in diamond can be deduced from the diamond laws.

Proof. By induction on the number of variables.

Let $F = G$ be an identity with N variables.

(Initial step.) If $N = 0$, then $F = G$ is an arithmetic equation. Since the diamond laws imply diamond's truth tables, $F = G$ follows from those axioms.

(Induction step.) Suppose that all $N - 1$ variable identities in diamond are provable from the diamond laws. Let $F(x)$ be F considered as an expression in its Nth variable x. Then:

$$F(x) = (F(\text{t}) \wedge x) \vee (F(\text{f}) \wedge {\sim}x) \vee \text{M}(F(\text{i}), \text{d}x, F(\text{j}))$$

is provable by the Differential Normal Form theorem. So is:

$$G(x) = (G(\text{t}) \wedge x) \vee (G(\text{f}) \wedge {\sim}x) \vee \text{M}(G(\text{i}), \text{d}x, G(\text{j})).$$

By the induction hypothesis, these are provable from the diamond laws:

$$F(\text{t}) = G(\text{t}); \quad F(\text{i}) = G(\text{i}); \quad F(\text{j}) = G(\text{j}); \quad F(\text{f}) = G(\text{f}).$$

Therefore we get this sequence of provable identities:

$$F(x) = (F(\text{t}) \wedge x) \vee (F(\text{f}) \wedge {\sim}x) \vee \text{M}(F(\text{i}), \text{d}x, F(\text{j}))$$
$$= (G(\text{t}) \wedge x) \vee (G(\text{f}) \wedge {\sim}x) \vee \text{M}(G(\text{i}), \text{d}x, G(\text{j}))$$
$$= G(x).$$

This concludes the induction proof. Therefore any equational identity in diamond is provable from the diamond laws.   QED

A note on feasibility. The above proof that $F = G$ was only three equations long; but it is only a link in a recursive chain. A complete proof requires proofs that $F(i) = G(i)$, $F(t) = G(t)$, $F(f) = G(f)$, and $F(j) = G(j)$. Therefore any complete proof that $F = G$, if these expressions have n variables, will be about 4^n steps long; no faster than proof by full-table look-up! Thus, though the diamond axioms are deductively complete, they may fail to be *feasibly* complete. Is there a polynomial-time algorithm that can check the validity of a general diamond equation? Students of feasibility will recognize this as a variant of the Boolean Consistency Problem, and therefore NP-complete.

Chapter 4

Self-Reference

Re-Entrance and Fixedpoints
Phase Order
The Outer Fixedpoints

A. Re-Entrance and Fixedpoints

Consider the Liar Paradox as a Brownian form:

$$L \quad = \quad \overline{L \;|}$$

This form contains itself. That can be represented via re-entrance, thus:

$$L \quad = \quad$$

Let re-entrance permit any mark within a Brownian form to extend a tendril to a distant space, where its endpoint shall be deemed enclosed. Thus curl sends a tendril into itself. Other re-entrant expressions include:

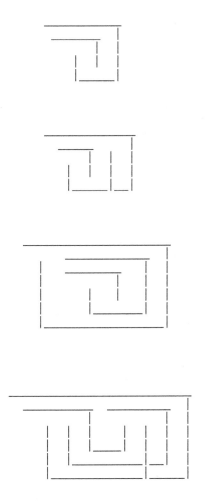

Self-reference can be expressed as a re-entrant Brownian form, as a switching circuit, as a vector of forms, as an indexed list, and as a harmonic fixedpoint. For example:

Brownian form

$$\cong$$

Switching Circuit
(triangles = "not" gates)

$$\cong$$

Brownian Form Vector

$$\cong$$

$$A = [[A]_B]_A$$

Indexed List

$$\cong$$

$$(A, B) = (\text{not } B, \text{ not } A)$$

Harmonic Fixedpoint

B.   Phase Order

Now let's define "phase order":

$$x \preceq y \quad \textbf{iff} \quad x \text{ min } y = x \quad \textbf{iff} \quad x \text{ max } y = y$$

```
                t
           ≼         ≼
    i          ≼          j
           ≼         ≼
                f
```

This structure is a lattice; it has a mutually distributive minimum and maximum.

Theorem. min *is the minimum operator for* $\preceq$;

$(X \text{ min } Y) \preceq X$; $(X \text{ min } Y) \preceq Y$;
and $Z \preceq (X \text{ min } Y)$, *if* $Z \preceq X$ *and* $Z \preceq Y$
Also: max *is the maximum operator for* $\preceq$;
$X \preceq (X \text{ max } Y)$; $Y \preceq (X \text{ max } Y)$;
and $(X \text{ max } Y) \preceq Z$, *if* $X \preceq Z$ *and* $Y \preceq Z$

Proof. $X \text{ min}(X \text{ min } Y) = (X \text{ min } X) \text{ min } Y = X \text{ min } Y$
ergo $(X \text{ min } Y) \preceq X$; similarly, $(X \text{ min } Y) \preceq Y$.
If $Z \preceq X$ and $Z \preceq Y$ then $Z \text{ min } X = Z$; also $Z \text{ min } Y = Z$;
Therefore $Z \text{ min}(X \text{ min } Y) = (Z \text{ min } X) \text{ min } Y) = Z \text{ min } Y = Z$.
Therefore $Z \preceq (X \text{ min } Y)$ if $Z \preceq X$ and $Z \preceq Y$.
Thus $(X \text{ min } Y)$ is the rightmost element left of both X and Y.
Similarly, $(X \text{ max } Y)$ is the leftmost element right of both X and Y.   QED

Theorem. $\preceq$ *is transitive and antisymmetric*:

$$a \preceq b \quad \text{and} \quad b \preceq c \quad \text{implies } a \preceq c,$$
$$a \preceq b \quad \text{and} \quad b \preceq a \quad \text{implies } a = b.$$

Proof. $a \preceq b$ and $b \preceq c$ implies

$$a \min b = a; \ \ b \min c = b; \text{ so}$$

$$a \min c = (a \min b) \min c = a \min(b \min c)$$
$$= a \min b = a; \text{ therefore } a \preceq c. \qquad \text{QED}$$

$a \preceq b$ and $b \preceq a$ implies

$$a \min b = a; \ \ a \min b = b; \text{ so } a = b. \qquad \text{QED}$$

Theorem. $\preceq$ *is preserved by disjunction and conjunction*:

$$a \preceq b \quad \text{implies} \quad a \text{ or } c \preceq b \text{ or } c$$
$$\text{and } a \text{ and } c \preceq b \text{ and } c$$

Proof. $a \preceq b$ implies $a \min b = a$; so

$$(a \text{ or } c) \min(b \text{ or } c) = (a \min b) \text{ or } c = a \text{ or } c;$$

so $(a \text{ or } c) \preceq (b \text{ or } c)$.

Similarly $(a \text{ and } c) \preceq (b \text{ and } c)$. \qquad QED

Theorem. $\preceq$ *is preserved by negation*:

$a \preceq b$ *implies* $\sim a \preceq \sim b$.

Proof. $a \preceq b$ implies $a \min b = a$; so

$$\sim a \min \sim b = \sim(a \min b) = \text{not } a$$

so $\sim a \preceq \sim b$. QED

Theorem. $\preceq$ *is preserved by any harmonic function*:

$a \preceq b$ *implies* $F(a) \preceq F(b)$.

This follows by induction from the previous two results.

Theorem. *For any harmonic f;*

$f(x \max y) \succeq f(x) \max f(y),$

$f(x \min y) \preceq f(x) \min f(y).$

Proof. By lattice properties.

$x \max y \succeq x; \ x \max y \succeq y$

ergo $f(x \max y) \succeq f(x)$

and $f(x \max y) \succeq f(y);$

so by definition of the max operator

$f(x \max y) \succeq f(y) \max f(y).$

We get the other half of the theorem the same way.

$f(x \min y) \preceq f(y) \min f(y).$ QED

These inequalities can be strict; for instance:

$$\text{dt min df} = f; \quad \text{yet d(t min f)} = i,$$

$$\text{Dt min Df} = t; \quad \text{yet D(t min f)} = i,$$

$$\text{dt max df} = f; \quad \text{yet d(t max f)} = j,$$

$$\text{Dt max Df} = t; \quad \text{yet D(t max f)} = j.$$

Now we extend $\preceq$ to ordered form vectors:

$$\mathbf{x} = (x_1, x_2, x_3, \ldots, x_n)$$

$$\mathbf{x} \preceq \mathbf{y} \quad \text{if and only if } (x_i \succeq y_i) \quad \text{for all } i.$$

Theorem. $\preceq$ *has "limited chains", with limit 2N.*

That is, if $\mathbf{x}_n$ is an ordered chain of finite form vectors; $\mathbf{x}_1 \preceq \mathbf{x}_2 \preceq \mathbf{x}_3 \ldots$, or $\mathbf{x}_1 \succeq \mathbf{x}_2 \succeq \mathbf{x}_3 \ldots$, and if N is the dimension of these vectors, then for all $n > 2N$, $\mathbf{x}_n = \mathbf{x}_{2N}$.

Proof. Any given component of the $\mathbf{x}$'s can move at most two steps before ending up at i or at j; at that point that component stops moving. For N components, this implies at most $2N$ steps in an ordered chain before it stops moving. QED

Given any harmonic function $\mathbf{f}(\mathbf{x})$, define

a *left seed* for $\mathbf{f}$ is any vector $\mathbf{a}$ such that $\mathbf{f}(\mathbf{a}) \preceq \mathbf{a}$;

a *right seed* for $\mathbf{f}$ is any vector $\mathbf{a}$ such that $\mathbf{a} \preceq \mathbf{f}(\mathbf{a})$.

a *fixedpoint* for $\mathbf{f}$ is any vector $\mathbf{a}$ such that $\mathbf{a} = \mathbf{f}(\mathbf{a})$.

A vector is a fixedpoint if and only if it is both a left seed and a right seed.

Left seeds generate fixedpoints, thus:

If $\mathbf{a}$ is a left seed for $\mathbf{f}$, then $\mathbf{f}(\mathbf{a}) \preceq \mathbf{a}$. Since $\mathbf{f}$ is harmonic, it preserves order; so $\mathbf{f}^2(\mathbf{a}) \succeq \mathbf{f}(\mathbf{a})$; and $\mathbf{f}^3(\mathbf{a}) \preceq \mathbf{f}^2(\mathbf{a})$; and so on:

$$\mathbf{a} \succeq \mathbf{f}(\mathbf{a}) \succeq \mathbf{f}^2(\mathbf{a}) \succeq \mathbf{f}^3(\mathbf{a}) \succeq \mathbf{f}^4(\mathbf{a}) \succeq \cdots .$$

Since diamond has limited chains, this descending sequence must reach its lower bound within $2n$ steps, if n is the number of components of $\mathbf{f}$. Therefore $\mathbf{f}^{2n}(\mathbf{a})$ is a *fixedpoint* for $\mathbf{f}$:

$$\mathbf{f}(\mathbf{f}^{2n}(\mathbf{a})) = \mathbf{f}^{2n}(\mathbf{a}).$$

This is the greatest fixedpoint left of $\mathbf{a}$.

Left seeds grow leftwards towards fixedpoints.

Similarly, right seeds grow rightwards towards fixedpoints:

$$\mathbf{a} \preceq \mathbf{f}(\mathbf{a}) \preceq \mathbf{f}^2(\mathbf{a}) \preceq \mathbf{f}^3(\mathbf{a}) \preceq \mathbf{f}^4(\mathbf{a}) \preceq \cdots \preceq \mathbf{f}^{2n}(\mathbf{a}) = \text{fixedpoint}$$

$\mathbf{f}^{2n}(\mathbf{a})$ is the leftmost fixedpoint right of the right seed $\mathbf{a}$.

All fixedpoints are both left and right seeds — of themselves.

C.   The Outer Fixedpoints

Now that we have self-referential forms, the question is; can we evaluate them in diamond logic? And if so, how?

It turns out that phase order permits us to do so in general. For any harmonic function $\mathbf{F}(\mathbf{x})$, we have the following:

The Self-Reference Theorem. *Any self-referential harmonic system has a fixedpoint*:

$$\mathbf{F}(\mathbf{x}) = \mathbf{x}.$$

Proof. Recall that all harmonic functions preserve order.

$\mathbf{i}$ is the leftmost set of values, hence this holds:

$$\mathbf{i} \preceq \mathbf{F}(\mathbf{i}).$$

Therefore, $\mathbf{i}$ is a right seed for $\mathbf{F}$:

$$\mathbf{i} \preceq \mathbf{F}(\mathbf{i}) \preceq \mathbf{F}^2(\mathbf{i}) \preceq \mathbf{F}^3(\mathbf{i}) \preceq \cdots \mathbf{F}^{2n}(\mathbf{i}) = \mathbf{F}(\mathbf{F}^{2n}(\mathbf{i}))$$

$\mathbf{i}$ generates the "leftmost" fixedpoint.   QED

Similarly, $\mathbf{j}$ generates the "rightmost" fixedpoint:

$$\mathbf{F}(\mathbf{F}^{2n}(\mathbf{j})) = \mathbf{F}^{2n}(\mathbf{j}).$$

All other fixedpoints lie between the two outermost:

$$\mathbf{F}^{2n}(\mathbf{i}) \preceq \mathbf{x} = \mathbf{F}(\mathbf{x}) \preceq \mathbf{F}^{2n}(\mathbf{j}).$$

I call this process "productio ex absurdo"; literally, production from the absurd; in contrast to "reduction to the absurd", boolean logic's refutation method. Diamond logic begins where boolean logic ends.

To see productio ex absurdo in action, consider this system:

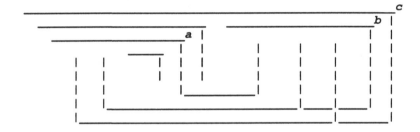

$$C = [[[BC[]]_A][ABC]_B]_C :$$

Iterate this system from curl:

The leftmost fixedpoint is:

$A =$ void, $B =$ curl, $C =$ void.

Iterating from uncurl yields the rightmost fixedpoint:

$A =$ void, $B =$ uncurl, $C =$ void.

All fixedpoints are between the outer fixedpoints; therefore $A = C =$ void; therefore $B =$ cross B; therefore $B =$ curl or uncurl. Thus the outer fixedpoints are the only ones.

Now consider the system:

Alan: "The key is in the drawer".
Bob: "If I'm right, then Alan is right".
Carl: "Alan is right or wrong".
Dan: "Alan is right, but Carl is right".
Eli: "Carl is right, but Alan is right".
Fred: "Dan is right or wrong".
Gary: "Eli is right or wrong".
Harry: "Fred and Gary are both right".

If two-valued logic were in control of this situation, then Bob's Santa sentence would make Alan right; and everyone else would be right too. Yet when you look in the drawer, the key isn't there!

The equations are:

$A = $ false,
$B = $ if B then A,
$C = A$ or not A,
$D = A$ but C,
$E = C$ but A,
$F = D$ or not D,
$G = E$ or not E,
$H = F$ and G.

If we iterate this system from default value **i**, we get:

$$(A, B, C, D, E, F, G, H) = (i, i, i, i, i, i, i, i)$$
$$\rightarrow (f, i, i, i, i, i, i, i)$$
$$\rightarrow (f, i, t, f, i, i, i, i)$$
$$\rightarrow (f, i, t, j, i, t, i, i)$$
$$\rightarrow (f, i, t, j, i, j, i, i)$$
$$\rightarrow (f, i, t, j, i, j, i, f)$$
$$= \text{fixedpoint.}$$

If we had started from default value j, we would have gotten the rightmost fixedpoint $(f, \mathbf{J}, t, j, i, j, i, f)$. As above, these are the only two fixedpoints.

Can the process take all of $2n$ steps? Yes. Consider the function:

$$f(x) = dx/t.$$

It takes two steps to go from i to the fixedpoint j;

$$i \rightarrow t \rightarrow j.$$

The following system takes $2n$ steps to go from **i** to the fixedpoint **j**:

$$x_1 = dx_1/t$$
$$x_2 = x_1 \min \sim x_1 \min dx_2/t$$
$$x_3 = x_2 \min \sim x_2 \min dx_3/t$$
$$\vdots$$
$$x_n = x_{n-1} \min \sim x_{n-1} \min dx_n/t$$

```
iii...i → tii...i → jii...i → jti...i
    → jji...i → jjt...i → ... → jjj...j
```

If the iterated function contains no reference to i or to j, then the process takes at most n steps, as in these systems:

$x_1 = \mathrm{f}; \ x_2 = Dx_1; \ x_3 = dx_2; \ x_4 = Dx_3$

4 steps:

```
iiii → fiii → ftii → ftfi → ftft
jjjj → fjjj → ftjj → ftfj → ftft
```

$x_1 = \mathrm{t}; \ x_2 = dx_1; \ x_3 = Dx_2; \ x_4 = dx_3; \ x_5 = Dx_4$

5 steps:

```
iiiii → tiiii → tfiii → tftii → tftfi → tftft
jjjjj → tjjjj → tfjjj → tftjj → tftfj → tftft
```

Chapter 5

Fixedpoint Lattices

Relative Lattices
Seeds and Spirals
Shared Fixedpoints
Examples

A.   Relative Lattices

Any harmonic function $\mathbf{F}(\mathbf{x})$ has the outer fixedpoints: $\mathbf{F}^n(\mathbf{i})$, $\mathbf{F}^n(\mathbf{j})$, the *leftmost* and *rightmost* fixedpoints. But often this is not all. In general, $\mathbf{F}$ has an entire *lattice* of fixedpoints.

Theorem.   *If* $\mathbf{a}$ *and* $\mathbf{b}$ *are fixedpoints for a harmonic function* F, *then these fixedpoints exist*:

$\mathbf{a} \min_F \mathbf{b} =$ *the rightmost fixedpoint left of both* $\mathbf{a}$ *and* $\mathbf{b}$
$\qquad = \mathbf{F}^{2n}(\mathbf{a} \min \mathbf{b})$,

$\mathbf{a} \max_F \mathbf{b} =$ *the leftmost fixedpoint right of both* $\mathbf{a}$ *and* $\mathbf{b}$
$\qquad = \mathbf{F}^{2n}(\mathbf{a} \max \mathbf{b})$.

Proof.   Let $\mathbf{a}$ and $\mathbf{b}$ be fixedpoints, and let $\mathbf{c}$ be any fixedpoint such that $\mathbf{c} \preceq \mathbf{a}$ and $\mathbf{c} \preceq \mathbf{b}$. Then $(\mathbf{a} \min \mathbf{b}) \preceq \mathbf{c}$; so
$$(\mathbf{a} \min \mathbf{b}) = \mathbf{F}(\mathbf{a}) \min \mathbf{F}(\mathbf{b}) \succeq \mathbf{F}(\mathbf{a} \min \mathbf{b}) \succeq \mathbf{F}(\mathbf{c}) = \mathbf{c}.$$

Ergo $(\mathbf{a} \min \mathbf{b})$ is a left seed greater than $\mathbf{c}$:

$$(\mathbf{a} \min \mathbf{b}) \succeq \mathbf{F}(\mathbf{a} \min \mathbf{b}) \succeq \mathbf{F}^2(\mathbf{a} \min \mathbf{b}) \succeq \cdots \succeq \mathbf{F}^{2n}(\mathbf{a} \min \mathbf{b})$$
$$= \mathbf{F}(\mathbf{F}^{2n}(\mathbf{a} \min \mathbf{b})) \succeq \mathbf{c}.$$

Therefore $\mathbf{F}^{2n}(\mathbf{a} \min \mathbf{b})$ is a fixedpoint left of $\mathbf{a}$ and of $\mathbf{b}$, and is moreover the rightmost such fixedpoint.

Thus, $\mathbf{F}^{2n}(\mathbf{a} \min \mathbf{b}) = \mathbf{a} \min_F \mathbf{b}$.   QED

Similarly, $\mathbf{F}^{2n}(\mathbf{a} \max \mathbf{b}) = \mathbf{a} \max_F \mathbf{b}$.   QED

For instance, consider the following Brownian form:

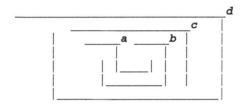

This is equivalent to this bracket-form system:

$$a = [b]; \quad b = [a]; \quad c = [ab]; \quad d = [cd].$$

In the standard interpretation, (a, b, c, d) is a fixedpoint for:

$$F(a, b, c, d) = (\sim b, \sim a, \sim(a \vee b), \sim(c \vee d))$$

In the nand-gate interpretation:

$$d = \sim(d \wedge c) = (\sim d \vee \sim c) = (d \Rightarrow da).$$

Sentence d says, "If I'm not mistaken, then sentence A is both true and false": a Lower Differential Santa Sentence!

In the nor-gate interpretation:

$d = Da - d$;

Sentence d says, "A is true or false, and I am a liar".
An Upper Differential Grinch!

F has this fixedpoint lattice:

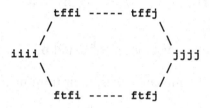

If we seek $\min_F$ of tffj and ftfj, then we must take their minimum, then apply F three times:

(tffj min ftfj) $=$ iifj $\to$ iiij $\to$ iiit $\to$ iiii

B.   Seeds and Spirals

We can generalize the preceding results to "seeds".

Theorem. *The minimum of two left seeds is a left seed.*

Proof. Let $\mathbf{c} = \mathbf{a}$ min $\mathbf{b}$, where $\mathbf{a}$ and $\mathbf{b}$ are left seeds.

Then $\mathbf{c} \preceq \mathbf{a}$, and $\mathbf{c} \preceq \mathbf{b}$, and $\mathbf{c}$ is the rightmost such vector.

Therefore $\mathbf{f}(\mathbf{c}) \preceq \mathbf{f}(\mathbf{a}) \preceq \mathbf{a}$; $\mathbf{f}(\mathbf{c}) \preceq \mathbf{f}(\mathbf{b}) \preceq \mathbf{b}$;

therefore $\mathbf{f}(\mathbf{c}) \preceq \mathbf{c}$,

since $\mathbf{c}$ is the rightmost vector left of $\mathbf{a}$ and $\mathbf{b}$.   QED

Since all fixedpoints are seeds, their minima are left seeds.

Theorem. *The minimum of left seeds generates the minimum of the fixedpoints in the relative lattice*:

$\mathbf{a}$ min $\mathbf{b}$ generates $\mathbf{f}^{2n}(\mathbf{a})$ $\min_f \mathbf{f}^{2n}(\mathbf{b})$,   if $\mathbf{a}$ and $\mathbf{b}$ are left seeds.

Proof. Let $\mathbf{z} = \mathbf{a}$ min $\mathbf{b}$, two left seeds. As noted above, $\mathbf{f}(\mathbf{z}) \preceq \mathbf{z}$; $\mathbf{z}$ is a left seed; moreover, $\mathbf{z}$ is the rightmost vector left of $\mathbf{a}$ and of $\mathbf{b}$. $\mathbf{f}^{2n}(\mathbf{z})$ is a fixedpoint; it's left of $\mathbf{f}^{2n}(\mathbf{a})$ and of $\mathbf{f}^{2n}(\mathbf{b})$, because $\mathbf{z}$ is left of $\mathbf{a}$ and of $\mathbf{b}$.

If a fixedpoint $\mathbf{c}$ is to the left of $\mathbf{f}^{2n}(\mathbf{a})$ and of $\mathbf{f}^{2n}(\mathbf{b})$, then $\mathbf{c}$ is to the left of $\mathbf{a}$ and of $\mathbf{b}$:

$$\mathbf{c} \preceq \mathbf{f}^{2n}(\mathbf{a}) \preceq \mathbf{a}; \quad \mathbf{c} \preceq \mathbf{f}^{2n}(\mathbf{b}) \preceq \mathbf{b};$$

$$\mathbf{c} \preceq \mathbf{a} \quad \text{and} \quad \mathbf{c} \preceq \mathbf{b} \quad \text{and} \quad \mathbf{z} = \mathbf{a} \min \mathbf{b};$$

therefore $\mathbf{c} \preceq \mathbf{z}$;

therefore $\mathbf{c} \preceq \mathbf{f}(\mathbf{z}) \preceq \mathbf{z}$;

therefore $\mathbf{c} \preceq \mathbf{f}^{2n}(\mathbf{z}) \preceq \cdots \preceq \mathbf{f}^{2}(\mathbf{z}) \preceq \mathbf{f}(\mathbf{z}) \preceq \mathbf{z}$.

$\mathbf{f}^{2n}(\mathbf{z})$ is a fixedpoint left of $\mathbf{f}^{2n}(\mathbf{a})$ and of $\mathbf{f}^{2n}(\mathbf{b})$, and it is the rightmost such fixedpoint. Therefore $\mathbf{z}$ generates the minimum in the relative lattice:

$$\mathbf{f}^{2n}(\mathbf{a} \min \mathbf{b}) = \mathbf{f}^{2n}(\mathbf{a}) \min_f \mathbf{f}^{2n}(\mathbf{b}). \qquad \text{QED}$$

In summary: the minimum of left seeds is a left seed, one which generates the relative minimum of the generated fixedpoints. Since any fixedpoint is a left seed, it follows that the minimum of fixedpoints is a left seed generating the relative minimum of the fixedpoints:

$$\mathbf{f}^{2n}(\mathbf{a} \min \mathbf{b}) = \mathbf{a} \min_f \mathbf{b}, \quad \text{if } \mathbf{a} \text{ and } \mathbf{b} \text{ are fixedpoints.}$$

And dually: the maximum of right seeds is a right seed, which generates the relative maximum of the generated fixedpoints. Since any fixedpoint is a right seed, the maximum of two fixedpoints is a right seed generating the relative maximum of the fixedpoints:

$$\mathbf{f}^{2n}(\mathbf{a} \max \mathbf{b}) = \mathbf{a} \max_f \mathbf{b}, \quad \text{if } \mathbf{a} \text{ and } \mathbf{b} \text{ are fixedpoints.}$$

We can generate a seed from a "spiral coil".

Definition. A *Left Spiral* is a function iteration sequence $\mathbf{v}_i = F^i(\mathbf{v}_0)$ such that, for some N and K, $\mathbf{v}_{K+N} \preceq \mathbf{v}_K$.
A *Left Spiral Coil* $= \{\mathbf{v}_K, \mathbf{v}_{K+1}, \ldots, \mathbf{v}_{K+N-1}\}$.

For all left spirals, these relations hold:

$$\mathbf{v}_K \succeq \mathbf{v}_{K+N} \succeq \cdots \succeq \mathbf{v}_{K+nN} = \mathbf{v}_{K+nN+N} = \cdots$$

$$\mathbf{v}_{K+1} \succeq \mathbf{v}_{K+N+1} \succeq \cdots \succeq \mathbf{v}_{K+nN+1} = \mathbf{v}_{K+nN+N+1} = \cdots$$

$$\mathbf{v}_{K+2} \succeq \mathbf{v}_{K+N+2} \succeq \cdots \succeq \mathbf{v}_{K+nN+2} = \mathbf{v}_{K+nN+N+2} = \cdots$$

$$\vdots$$

$$\mathbf{v}_{K+N-1} \succeq \mathbf{v}_{K+2N-1} \succeq \cdots \succeq \mathbf{v}_{K+nN+N-1} = \mathbf{v}_{K+nN+2N-1} = \cdots,$$

where n is the dimension of the vectors.

The spiral coils leftwards until it reaches a limit cycle.

Spiral Theorem. *The minimum of a left spiral coil is a left seed.*

Proof.

$$F(\mathbf{v}_K \min \mathbf{v}_{K+1} \min \ldots \min \mathbf{v}_{K+N-1})$$

$$\preceq F(\mathbf{v}_K) \min F(\mathbf{v}_{K+1}) \min \ldots \min F(\mathbf{v}_{K+N-1})$$

$$= \mathbf{v}_{K+1} \min \mathbf{v}_{K+2} \min \ldots \min \mathbf{v}_{K+N-1} \min \mathbf{v}_{K+N}$$

$$\preceq \mathbf{v}_{K+1} \min \mathbf{v}_{K+2} \min \ldots \min \mathbf{v}_{K+N-1} \min \mathbf{v}_K$$

$$= \mathbf{v}_K \min \mathbf{v}_{K+1} \min \ldots \min \mathbf{v}_{K+N-1}. \qquad \text{QED}$$

The fixedpoint that grows from this left seed is the rightmost fixedpoint left of the spiral's limit cycle; a "wave-bracketing fixedpoint".

Of course there are similar results for right spiral coils, etc.

C. Shared Fixedpoints

More than one function can share a fixedpoint. For instance:

Theorem. *If* $F(x)$ *and* $G(x)$ *are harmonic functions and* $F(G(x)) = G(F(x))$ (F *and* G *commute). Then* F *and* G *share a nonempty lattice of fixedpoints*:

$$F(x) = x; \quad G(x) = x \quad \text{for all } x \text{ in } L_{FG}.$$

Proof. We have proved that the harmonic function G has a lattice of fixedpoints; $G(x) = x$ for all x in L_G.

But since F commutes with G,

$$G(F(x)) = F(G(x)) = F(x) \quad \text{for all } x \text{ in } L_G.$$

that is, F sends fixedpoints of G to fixedpoints of G.

F sends L_G to itself. What's more, F preserves order in diamond; therefore F preserves order in L_G.

Therefore F is an order-preserving function from L_G to itself. That fact, plus lattice arguments like those in previous sections, will prove that F has fixedpoints in a nonempty lattice L_{FG} of L_G:

$$F(x) = x \quad \text{and} \quad G(x) = x \quad \text{for every element } x \text{ of } L_{FG}.$$

QED

L_{FG} is a lattice of shared fixedpoints.

Its least element is $\mathbf{F}^{2n}(\mathbf{G}^{2n}(\mathbf{i}))$;

its greatest element is $\mathbf{F}^{2n}(\mathbf{G}^{2n}(\mathbf{j}))$;

and its relative minimum operator is $\mathbf{F}^{2n}(\mathbf{G}^{2n}(\mathbf{a} \min \mathbf{b}))$.

and its relative maximum operator is $\mathbf{F}^{2n}(\mathbf{G}^{2n}(\mathbf{a} \max \mathbf{b}))$.

These results can be extended to N functions:

If $\mathbf{F}_1, \mathbf{F}_2, \ldots, \mathbf{F}_N$ are N commuting harmonic functions, then they share a semilattice of fixedpoints:

$$\mathbf{F}_i(\mathbf{x}) = \mathbf{x} \quad \text{for all } i \text{ between 1 and } N, \text{ and all } \mathbf{x} \text{ in } L.$$

Its least element is $\mathbf{F}_1^{2n}(\mathbf{F}_2^{2n}(\cdots(\mathbf{F}_N^{2n}(\mathbf{i}))\cdots))$;

its greatest element is $\mathbf{F}_1^{2n}(\mathbf{F}_2^{2n}(\cdots(\mathbf{F}_N^{2n}(\mathbf{j}))\cdots))$;

its relative minimum operator is

$$\mathbf{F}_1^{2n}(\mathbf{F}_2^{2n}(\cdots(\mathbf{F}_N^{2n}(\mathbf{a} \min \mathbf{b}))\cdots));$$

and its relative maximum operator is

$$\mathbf{F}_1^{2n}(\mathbf{F}_2^{2n}(\cdots(\mathbf{F}_N^{2n}(\mathbf{a} \max \mathbf{b}))\cdots)).$$

D. Examples

Consider the liar paradox:

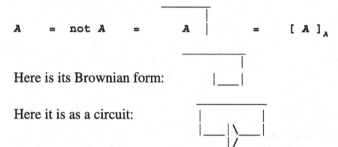

$$A \quad = \quad \text{not } A \quad = \quad A \; \overline{} \; | \quad = \quad [\, A \,]_A$$

Here is its Brownian form: $\overline{}|$ $|\_\_|$

Here it is as a circuit:

```
    _____
   |        |
   |__|\ ___|
      |/
```

Here is its fixedpoint lattice: i ----------- j

Now consider Tweedle's Quarrel:

Tweedledee: "Tweedledum is a liar".
Tweedledum: "Tweedledee is a liar".

$$E \quad = \quad \overline{U} \, | \qquad \qquad E = [\,[\,E\,]_U\,]_E$$

$$U \quad = \quad \overline{E} \, |$$

```
    _____
          | |
     | |  | |
     |_____|
```

Its circuit is:

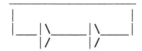

This "toggle's" lattice is:

Consider the following statement:

"This statement is both true and false".

It resolves to this system, the "duck": $B = [[B]_A B]_B$

$$A \quad = \quad \overline{B} \mid \qquad \overline{\qquad\qquad a}^{b}$$

$$B \quad = \quad \overline{AB} \mid$$

I gave the "Duck" that name because of the appearance of its circuit:

This is equivalent to the fixedpoint:

$$B = (B \wedge \sim B) = dB; \quad \text{a differential of itself!}$$

Here is its lattice: ii ----- tf ----- jj

This is the "truck": $C = [[[A]_A C]_B]_C$

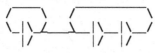

$$
\begin{aligned}
A &= \overline{A} \,| \\
B &= \overline{CA}\,| \\
C &= \overline{B} \,|
\end{aligned}
$$

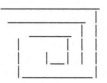

It has this lattice: iii-----ift-----jft-----jjj

This jolly-looking form: $C = [[[BC]_A[A]_B][C]]_C$

$$
\begin{aligned}
A &= \overline{B\ C}\,| \\
B &= \overline{A}\,| \\
C &= \overline{A\ B}\,|\ \overline{C}\,|\,|
\end{aligned}
$$

has this lattice:

```
                     ftt
             fti            ftj
     iii             ftf            jjj
             iif            jjf
                     tff
```

The "rabbit": $D = [[[[B]_A C]_B D]_C]_D$

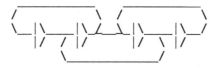

$$
\begin{aligned}
A &= B \\
B &= AC \\
C &= BD \\
D &= C
\end{aligned}
$$

has a similar lattice:

```
                        tftf
             tfii                tfjj
    iiii                tfft                jjjj
             iift                jjft
                        ftft
```

To create linear fixedpoint lattices of length $2n + 1$, use:

$$x_1 = Dx_1,$$

$$x_2 = Dx_2 \vee x_1,$$

$$x_3 = Dx_3 \vee x_2,$$

$$\vdots$$

$$x_n = Dx_n \vee x_{n-1}.$$

Its circuit is:

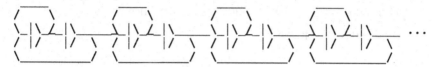

I call this circuit "the ducks".

For $n = 4$, we get the lattice:

`iiii - iiit - iitt - ittt - tttt - jttt - jjtt - jjjt - jjjj`

To create linear fixedpoint lattices of length $2n$, use:

$$x_1 = {\sim}x_1,$$

$$x_2 = Dx_2 \vee x_1,$$

$$x_3 = Dx_3 \vee x_2,$$

$$\vdots$$

$$x_n = Dx_n \vee x_{n-1}.$$

For $n = 4$, we get the lattice:

`iiii - iiit - iitt - ittt - jttt - jjtt - jjjt - jjjj`

This Brownian form: $c = [[[a]a]_a[ac]_b]_c$

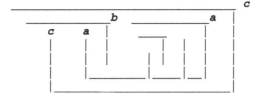

is equivalent to the bracket-form system:

$a = [a[a]]; \quad b = [ac]; \quad c = [ab].$

That is: $a = \mathrm{d}a; \quad b = a$ nor $c; \quad c = a$ nor b.

Its fixedpoint lattice is:

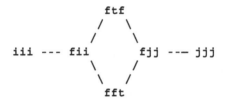

In general, the system

$a = \mathrm{d}a;$

$\mathbf{b} = M(a, \sim a, \mathbf{f}(\mathbf{b}))$

will have this fixedpoint lattice:

```
ii ---(  L  ) --- jj
```

where L is $\mathbf{f}$'s fixedpoint lattice.

This system:

$$a = [b]; \quad b = [a]; \quad c = [ad]; \quad d = [acd]; \quad e = [bf]; \quad f = [be]$$

has this fixedpoint lattice:

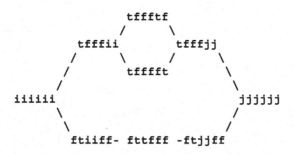

The "toggle" *ab* controls which subcircuit activates; the toggle *ef* or the "duck" *cd*. In general, the system

$$a = \sim b;$$

$$b = \sim a;$$

$$\mathbf{c} = (a \wedge \mathbf{f(c)}) \vee (b \wedge \mathbf{g(c)}) \vee da$$

will have this fixedpoint lattice:

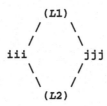

where *L*1 is **f**'s lattice, and where *L*2 is **g**'s lattice.

The "triplet" has this form: $C = [[BC]_A[CA]_B]_C$

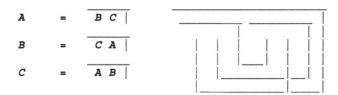

$$
\begin{array}{ccc}
A & = & \overline{B\ C}\ | \\
B & = & \overline{C\ A}\ | \\
C & = & \overline{A\ B}\ |
\end{array}
$$

The triplet has this circuit:

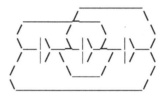

Its lattice is:

```
           tff
          /    \
         /      \
iii -- ftf -- jjj
         \      /
          \    /
           fft
```

Note that this lattice (called "M_3") is non-distributive:

$$a$$

$$0 \quad < \quad b \quad < \quad 1$$

$$c$$

$(a \max b) \min c = 1 \min c = c,$

$(a \min c) \max (b \min c) = 0 \max 0 = 0.$

On the other hand, it *is* "modular":

$x \leq z$ implies $x \max (y \min z) = (x \max y) \min z.$

It is a theorem of lattice theory that any non-distributive modular lattice contains M_3 as a sublattice.

In general, the system

$a = {\sim}(b \vee c); \quad b = {\sim}(c \vee a); \quad c = {\sim}(a \vee b);$

$\mathbf{d} = (a \wedge \mathbf{f}(\mathbf{d})) \vee (b \wedge \mathbf{g}(\mathbf{d})) \vee (c \wedge \mathbf{h}(\mathbf{d})) \vee (a \wedge b \wedge c).$

will have this fixedpoint lattice:

```
              (L₁)
             /    \
            /      \
    iiii--(L₂)--jjjj
            \      /
             \    /
              (L₃)
```

where L_1, L_2, and L_3 are the lattices for $\mathbf{f}$, $\mathbf{g}$, and $\mathbf{h}$.

The "ant" or "toggled buzzer", has the form $C = [[[B]_A]_B C]_C$:

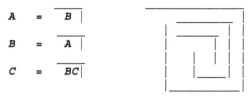

$$A = \overline{B} \mid$$

$$B = \overline{A} \mid$$

$$C = \overline{BC} \mid$$

The ant's lattice is:

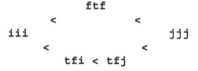

ftf

iii   jjj

tfi < tfj

Note that this lattice (called N_5) is non-distributive:

$$b$$

$$0 \qquad 1$$

$$a < c$$

$$(a \max b) \min c = 1 \min c = c,$$

$$(a \min c) \max (b \min c) = a \max 0 = a.$$

It is also non-modular:

$$a < c, \quad \text{but } a \max(b \min c) = a \max 0 = a;$$

$$\text{and } (a \max b) \min c = 1 \min c = c.$$

It is a theorem of lattice theory that any non-distributive non-modular lattice contains N_5 as a sublattice.

Note that the FTF state is the ant's only boolean state; all others contain paradox. Assuming that gate C is boolean forces gates A and B to be in the FT state only. The "ant" thus resembles the "Santa" statements of Chapter 1; both attempt to use the threat of paradox to force values otherwise free.

Here is the "goldfish":

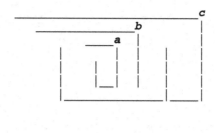

This is equivalent to this bracket-form system:

$a = [a];$ $b = [ac];$ $c = [bc] = [[ac]c].$

In the nand-gate interpretation:

$c = (c \Rightarrow (a \wedge c))$

$a = $ "I am a liar".

$c = $ "If I am honest, then we're both honest".

A Self and Buzzer Santa! It has this fixedpoint lattice:

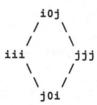

Note that when $b = 0$, this is due to Duality: [ij] = 0. The buzzer c seals itself off from the buzzer a.

Consider this Brownian form:

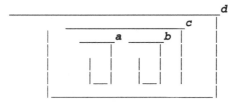

This is equivalent to this bracket-form system:

$a = [a];$ $b = [b];$ $c = [ab];$ $d = [cd].$

In the nand-gate interpretation:

$$d = \sim(d \wedge c) = \sim d \vee \sim c = (d \Rightarrow (a \vee b))$$

$A = $ "I am a liar".

$B = $ "I am a liar".

$D = $ "If I am honest, then they're both honest".

A Two-Buzzers Santa!

F has this fixedpoint lattice:

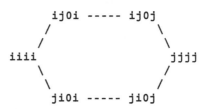

Note that when $C = 0$, this is due to Duality: [ij] = 0.

Now consider this Brownian form; "two ducks in a box":

$$C = [[a[a]]_a[b[b]]_bc]_c$$

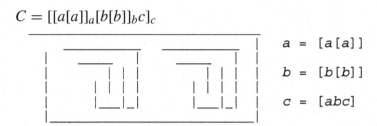

a = [a[a]]

b = [b[b]]

c = [abc]

In the nand interpretation, this is:

$a = Da =$ "I am honest or a liar".

$b = Db =$ "I am honest or a liar".

$c = a$ nand b nand $c =$ "One of us is a liar".

The fixedpoint lattice of this "Complementarity Santa" is:

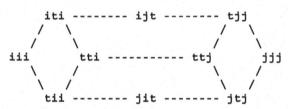

Note the fixedpoints ijt and jit; these are the only ones where C has a boolean value; but this is due to Complementarity, an anti-boolean axiom! Without those points, this lattice would be modular; but with them it contains N_5.

Consider this form; the "First Brownian Modulator":

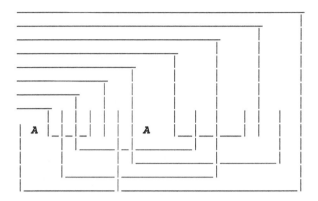

It is equivalent to the boundary-logic system:

A = input,

$B = [KA]$,

$C = [BD]$,

$D = [BE]$,

$E = [DF]$,

$F = [HA]$,

$G = [FE]$,

$H = [KC]$,

$K = [HG]$.

If we symbolize the marked state by "1", curl by "i", uncurl by "j" and unmarked by "0", then this system has these fixedpoints:

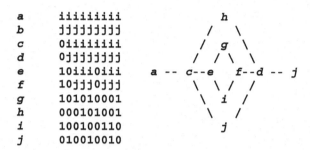

```
a       iiiiiiiii                        h
b       jjjjjjjjj                      /   \
c       0iiiiiiii                     /  g  \
d       0jjjjjjjj                    /  / \  \
e       10iii0iii          a -- c--e    f--d -- j
f       10jjj0jjj                    \  \ /  /
g       101010001                     \  i  /
h       000101001                      \   /
i       100100110                        j
j       010010010
```

George Spencer-Brown, in his *Laws of Form*, claims that this circuit "counts to two"; i.e. when *A* oscillates twice between marked and unmarked, *K* oscillates once. In Chapter 12, "How to Count to Two", we will 'diffract' this circuit, and reveal a simple rotor.

Chapter 6

Limit Logic

Limits
Limit Fixedpoints
The Halting Theorem

A. Limits

Diamond logic is continuous; it defines limit operators. These operators equal combinations of two more familiar limit operators; "infinity" and "cofinity":

$$\operatorname{Inf}(x_n) = (\text{All } N \geq 0)(\text{Exists } n \geq N)(x_n)$$
$$= (x_1 \vee x_2 \vee x_3 \vee x_4 \vee \ldots)$$
$$\wedge (x_2 \vee x_3 \vee x_4 \vee \ldots)$$
$$\wedge (x_3 \vee x_4 \vee \ldots)$$
$$\wedge (x_4 \vee \ldots)$$
$$\wedge \ldots$$

$$\operatorname{Cof}(x_n) = (\text{Exists } N \geq 0)(\text{All } n \geq N)(x_n)$$
$$= (x_1 \wedge x_2 \wedge x_3 \wedge x_4 \wedge \ldots)$$
$$\vee (x_2 \wedge x_3 \wedge x_4 \wedge \ldots)$$
$$\vee (x_3 \wedge x_4 \wedge \ldots)$$
$$\vee (x_4 \wedge \ldots)$$
$$\vee \ldots$$

Inf, the "infinity" quantifier, says that x_n is true infinitely often. Cof, the "cofinity" operator, says that x_n is false only finitely often. Obviously these are deeply implicated in the Paradox of Finitude.

Note that cofinity is a stricter condition; cofinite implies infinite, but not necessarily the reverse:

$$\text{Inf}(x_n) \vee \text{Cof}(x_n) = \text{Inf}(x_n);$$
$$\text{Inf}(x_n) \wedge \text{Cof}(x_n) = \text{Cof}(x_n).$$

Now define a "directed limit" via majorities:
$$\lim{}^a(x_n) = M(\text{Inf}(x_n), a, \text{Cof}(x_n)).$$

Note that:

$$\lim{}^f(x_n) = M(\text{Inf}(x_n), f, \text{Cof}(x_n)) = \text{Cof}(x_n),$$
$$\lim{}^t(x_n) = M(\text{Inf}(x_n), t, \text{Cof}(x_n)) = \text{Inf}(x_n).$$

The intermediate settings define "limit" operators;

$$
\begin{aligned}
\lim{}^-(x_n) = \quad & \lim{}^i(x_n) \\
= \quad & M(\text{Inf}(x_n), i, \text{Cof}(x_n)) \\
= \quad & \text{Inf}(x_n) \min \text{Cof}(x_n) \\
= \quad & \text{Inf}(x_n)/\text{Cof}(x_n) \\
= \quad & \text{Max}(N \geq 0)\text{Min}(n \geq N)(x_n) \\
= \quad & (x_1 \min x_2 \min x_3 \min x_4 \min \ldots) \\
& \max(x_2 \min x_3 \min x_4 \min \ldots) \\
& \quad \max(x_3 \min x_4 \min \ldots) \\
& \quad\quad \max(x_4 \min \ldots) \\
& \quad\quad\quad \max \ldots .
\end{aligned}
$$

$$\lim{}^+(x_n) = \lim{}^j(x_n)$$
$$= M(\mathrm{Inf}(x_n), j, \mathrm{Cof}(x_n))$$
$$= \mathrm{Inf}(x_n) \max \mathrm{Cof}(x_n)$$
$$= \mathrm{Cof}(x_n)/\mathrm{Inf}(x_n)$$
$$= \mathrm{Min}(N \geq 0)\mathrm{Max}(n \geq N)(x_n)$$
$$= (x_1 \max x_2 \max x_3 \max x_4 \max \ldots)$$
$$\min(x_2 \max x_3 \max x_4 \max \ldots)$$
$$\min(x_3 \max x_4 \max \ldots)$$
$$\min(x_4 \max \ldots)$$
$$\min \ldots.$$

Note that:

$$M(\lim{}^-(x_n), \mathrm{t}, \lim{}^+(x_n)) = \lim{}^-(x_n) \vee \lim{}^+(x_n)$$
$$= \mathrm{Inf}(x_n) = \lim{}^{\mathrm{t}}(x_n)$$
$$M(\lim{}^-(x_n), \mathrm{f}, \lim{}^+(x_n)) = \lim{}^-(x_n) \wedge \lim{}^+(x_n))$$
$$= \mathrm{Cof}(x_n) = \lim{}^{\mathrm{f}}(x_n)$$

and in general:

$$M(\lim{}^-(x_n), a, \lim{}^+(x_n)) = \lim{}^a(x_n).$$

Theorem.

$$(\lim{}^- x_{n+1}) = (\lim{}^- x_n),$$
$$(\lim{}^+ x_{n+1}) = (\lim{}^+ x_n).$$

This is true because Inf and Cof have that property. Inf and Cof are about the long run, not about the beginning.

Lim$^-$ is the *rightmost value left of cofinitely many* x_n's, and lim$^+$ is the *leftmost value right of cofinitely many* x_n's:

$\lim^- x_n \preceq x_n$, for all but finitely many N

and lim$^-$ is the rightmost such value;

$\lim^+ x_n \succeq x_n$, for all but finitely many N

and lim$^+$ is the leftmost such value.

Lim$^-$ and lim$^+$ are min, or max, respectively, of the *cofinal range* of x_n; the set of values that occur infinitely often:

$\lim^- \{x_n\} = \text{Min cofinal}\{x_n\}$,

$\lim^+ \{x_n\} = \text{Max cofinal}\{x_n\}$,

where $\text{cofinal}\{x_n\} = \{Y : x_n = Y \text{ for infinitely many } n\}$.

Theorem. *If F is a harmonic function, then*

$F(\lim^- x_n) \preceq \lim^- F(x_n)$;

$F(\lim^+ x_n) \succeq \lim^+ F(x_n)$.

Proof. We shall take the lim$^-$ case first.

$\lim^- x_n \preceq x_N$, for cofinitely many N.

Therefore: $F(\lim^- x_n) \preceq F(x_N)$, for cofinitely many N.

Therefore: $F(\lim^- x_n) \preceq \lim^- F(x_N)$,

since $\lim^- F(x_N)$ is the *rightmost* value left of cofinitely many $F(x_N)$'s!

The lim$^+$ case follows by symmetry.   QED

Here's another proof, by cofinality:

$$F(\lim{}^- x_n) = F(\text{Min cofinal}\{x_n\})$$
$$\preceq \text{Min } F(\text{cofinal}\{x_n\})$$
$$= \text{Min cofinal}\{F(x_n)\}$$
$$= \lim{}^- F(x_n). \qquad\qquad \text{QED}$$

These inequalities can be strict. For instance:

$$F(x) = \mathrm{d}x \quad \text{and} \quad x_n = \{\mathrm{t}, \mathrm{f}, \mathrm{t}, \mathrm{f}, \mathrm{t}, \mathrm{f}, \ldots\}:$$

$$\mathrm{d}(\lim{}^- \{\mathrm{t}, \mathrm{f}, \mathrm{t}, \mathrm{f}, \mathrm{t}, \mathrm{f}, \ldots\}) = \mathrm{d}i = i;$$
$$\lim{}^- \{\mathrm{d}\mathrm{t}, \mathrm{d}\mathrm{f}, \mathrm{d}\mathrm{t}, \mathrm{d}\mathrm{f}, \ldots\} = \lim{}^- \{\mathrm{f}, \mathrm{f}, \mathrm{f}, \mathrm{f}, \ldots\} = \mathrm{f},$$
$$\text{so } \mathrm{d}(\lim{}^- x_n) < \lim{}^- \mathrm{d}(x_n)$$

$$\mathrm{d}(\lim{}^+ \{\mathrm{t}, \mathrm{f}, \mathrm{t}, \mathrm{f}, \mathrm{t}, \mathrm{f}, \ldots\}) = \mathrm{d}j = j;$$
$$\lim{}^+ \{\mathrm{d}\mathrm{t}, \mathrm{d}\mathrm{f}, \mathrm{d}\mathrm{t}, \mathrm{d}\mathrm{f}, \ldots\} = \lim{}^+ \{\mathrm{t}, \mathrm{t}, \mathrm{t}, \mathrm{t}, \ldots\} = \mathrm{t},$$
$$\text{so } \mathrm{d}(\lim{}^+ x_n) > \lim{}^+ \mathrm{d}(x_n).$$

B.   Limit Fixedpoints

Fixedpoints can be found by transfinite induction on the limit operators. Recall that for all harmonic functions $\mathbf{F}$:

$$\mathbf{F}(\lim^- \mathbf{a}_n) \preceq \lim^- \mathbf{F}(\mathbf{a}_n),$$

$$\mathbf{F}(\lim^+ \mathbf{a}_n) \succeq \lim^+ \mathbf{F}(\mathbf{a}_n).$$

Given *any* set of initial values $\mathbf{s}_0$, then let

$$\mathbf{s}_\omega = \mathbf{F}^{\omega-}(\mathbf{s}_0) = \lim^- \mathbf{F}^n(\mathbf{s}_0).$$

So $\mathbf{F}(\mathbf{s}_\omega) = \mathbf{F}(\lim^- \mathbf{F}^n(\mathbf{s}_0))$

$$\preceq \lim^- \mathbf{F}(\mathbf{F}^n(\mathbf{s}_0))$$

$$= \lim^- \mathbf{F}^{n+1}(\mathbf{s}_0)$$

$$= \lim^- \mathbf{F}^n(\mathbf{s}_0)$$

$$= \mathbf{s}_\omega.$$

Therefore $\mathbf{F}^{\omega-}(\mathbf{s}_0) = \lim^-(\mathbf{F}^n(\mathbf{s}_0))$ is a left seed. It generates a fixedpoint:

$$\mathbf{s}_\omega \succeq \mathbf{F}(\mathbf{s}_\omega) \succeq \mathbf{F}^2(\mathbf{s}_\omega) \succeq \ldots \mathbf{s}_{2\omega} \; = \; \lim^- \mathbf{F}^n(\lim^-(\mathbf{F}^n(\mathbf{s}_0))).$$

$\mathbf{s}_{2\omega}$ is the limit of a descending sequence, and therefore also its minimum. If $\mathbf{F}$ has only finitely many components, then the descending sequence can only descend finitely many steps before coming to rest.

Thus, if $\mathbf{F}$ has finitely many components, then $\mathbf{s}_{2\omega}$ is a fixedpoint for $\mathbf{F}$: $\mathbf{F}(\mathbf{s}_{2\omega}) = \mathbf{s}_{2\omega}$.

Similarly, $\mathbf{F}^{\omega+}(\mathbf{s}_0) = \lim^+(\mathbf{F}^n(\mathbf{s}_0))$ is a right seed, which generates the fixedpoint $\mathbf{F}^{2\omega+}(\mathbf{s}_0) = \lim^+(\mathbf{F}^n(\lim^+(\mathbf{F}^n(\mathbf{s}_0))))$.

These seeds also generate minima and maxima in the relative lattice:

$$\lim{}^{-}(\mathbf{F}^{n}(\mathbf{F}^{\omega-}(x_0)\min\mathbf{F}^{\omega-}(y_0)))$$

$$=\mathbf{F}^{2\omega-}(x_0)\ \min{}_F\ \mathbf{F}^{2\omega-}(y_0),$$

$$\lim{}^{+}(\mathbf{F}^{n}(\mathbf{F}^{\omega+}(x_0)\max\mathbf{F}^{\omega+}(y_0)))$$

$$=\mathbf{F}^{2\omega+}(x_0)\ \max{}_F\ \mathbf{F}^{2\omega+}(y_0).$$

If **F** has *infinitely* many components, then we must continue the iteration through more limits. Let

$$s_{3\omega}=\lim{}^{-}(\mathbf{F}^{n}(s_{2\omega})),$$

$$s_{4\omega}=\lim{}^{-}(\mathbf{F}^{n}(s_{3\omega})),$$

$$\vdots$$

$$s_{\omega\omega}=\lim{}^{-}(s_{n\omega}),$$

$$\vdots$$

And so on through the higher ordinals. They keep drifting left; so at a high enough ordinal, we get a fixedpoint:

$$\mathbf{F}(s_\alpha)=s_\alpha.$$

Large cardinals imply "late" fixedpoints: self-reference with high complexity. Alas, the complexity is all in the syntax of the system, not its (mostly imaginary) content. Late fixedpoints are absurdly simple answers to absurdly complex questions.

C.   The Halting Theorem

Let

$$\mathbf{F}^{2\omega-}(\mathbf{s}_0) = \lim{}^-\mathbf{F}^n(\lim{}^-\mathbf{F}^n(\mathbf{s}_0))$$

and

$$\mathbf{F}^{2\omega+}(\mathbf{s}_0) = \lim{}^+\mathbf{F}^n(\lim{}^+\mathbf{F}^n(\mathbf{s}_0)).$$

These are the left and right fixedpoints generated from $\mathbf{s}_0$ by iterating $\mathbf{F}$ twice-infinity times. We can regard each of these as the output of a computation process whose input is $\mathbf{s}_0$ and whose program is $\mathbf{F}$. Diamond's computation theory is the same as its limit theory; output equals behavior "in the long run".

If $\mathbf{F}$ is n-dimensional, and if $\mathbf{F}^i(\mathbf{s}_0)$ is a cyclic pattern — that is, a wave — then $\lim{}^-(\mathbf{F}^i(\mathbf{s}_0))$ equals minimum over a cycle, and $\lim{}^+(\mathbf{F}^i(\mathbf{s}_0))$ equals maximum over a cycle. These yield the *wave-bracketing fixedpoints*:

$$\mathbf{F}^{2\omega-}(\mathbf{s}_0) = \lim{}^-\mathbf{F}^n(\lim{}^-\mathbf{F}^n(\mathbf{s}_0))$$
$$= \mathbf{F}^{2n}(\mathrm{Min}(\mathbf{F}^i(\mathbf{s}_0))),$$

where Min is taken over at least one cycle.

This is the rightmost fixedpoint left of cofinitely many $\mathbf{F}^i(\mathbf{s}_0)$.

$$\mathbf{F}^{2\omega+}(\mathbf{s}_0) = \lim{}^+\mathbf{F}^n(\lim{}^+\mathbf{F}^n(\mathbf{s}_0))$$
$$= \mathbf{F}^{2n}(\mathrm{Max}(\mathbf{F}^i(\mathbf{s}_0))),$$

where Max is taken over at least one cycle.

This is the leftmost fixedpoint right of cofinitely many $\mathbf{F}^i(\mathbf{s}_0)$.

Their existence implies this **Halting Theorem**.

If **F** has n components, then its limit fixedpoints equal:

$$\mathbf{F}^{2\omega-}(\mathbf{s}_0) = \mathbf{F}^{2n}\left(\min_{4^n < j < 2*4^n}(\mathbf{F}^j(\mathbf{s}_0))\right),$$

$$\mathbf{F}^{2\omega+}(\mathbf{s}_0) = \mathbf{F}^{2n}\left(\max_{4^n < j < 2*4^n}(\mathbf{F}^j(\mathbf{s}_0))\right).$$

This is because by 4^n steps, the system has run through all possible different states; so between 4^n and $2*4^n$ it will traverse at least one cycle, and thus generate a seed.

The minimum of stages 4^n to $2*4^n$, iterated $2n$ times more, yields a wave-bracketing fixedpoint, in $(2n + 2*4^n)$ steps.

In diamond logic, any computation with any input has an output; a wave-bracketing fixedpoint. However, some computations take exponential time to find their wave, and thus are nonfeasible.

Most of the logic fixedpoints in the last few chapters exist thanks to the default value; paradox. In diamond logic, paradox doesn't *refute* reasoning; it *grounds* reasoning.

Chapter 7
Paradox Resolved

A.   The Liar and the Anti-Diagonal

"This sentence is false"; is that, the pseudomenon, true or false? Yes but no! Or, if you prefer, no but yes. Before we had no solutions at all; now we have more than one!

Dear reader, I must confess to a sense of anticlimax in this resolution. So many logicians have treated paradox with respect bordering on terror; surely the solution can't be *that* simple? Well, yes it can be; for as you can see, yes it is!

Call an adjective "heterological" if and only if it does not apply to itself:

"A" is heterological = "A" is not A. Is "heterological" heterological?

"Heterological" is heterological = "Heterological" is not heterological.

Yes but no. (Or: no but yes.)

" 'Is false when quined' is false when quined"; is it true?

Yes but no.

B.   Russell's Paradox

Recall the definition of Russell's set R:

   $R = \{x | x \notin x\}$.
So in general $x \in R = x \notin x$
and therefore $R \in R = R \notin R$.
Therefore R is paradoxical. Does R exist?

In 2-valued logic, the answer must be "no"; yet there it is! In diamond logic, "R in R" equals i, or it equals j.

Recall also the "Short-Circuit Set": $S = \{x : S \notin S\}$.
X is not a variable in the definition of S, so S is constant-valued:
For all x, $(x \in S) = (S \notin S) = (S \in S)$.
All sets are paradox elements for S.

Russell's barber shaves all those — and only those — who do not shave themselves. Does the barber shave himself?

Yes but no; which can be realized several ways. For instance, the barber might only *partially* shave himself. Or, if there are *two* barbers in town, then each can shave each other, but not themselves; then the two of them, as a team, shave all those who do not shave themselves.

That village's watchmen watch all those, and only those, who do not watch themselves. But who watches the watchmen?

Answer: they shall watch each other, but not themselves. Thus honesty in government is truly imaginary!

If you were to ask that village's veterans about the Great War (a war to end all wars, and only those wars, which do not end themselves), then they will laugh at your quaint name for a conflict now known as World War I.

"Did the Great War end itself?" they will say, then scratch their heads. "Yes, it did; but no, it did not"!

That village's priest often ponders this theological riddle:

God is worshipped by all those, and only those, who do not worship themselves. Does God worship himself?

Answer: not this, not that. A mystery!

C.   Santa and the Grinch

If a young child were to proclaim:

"Santa Claus exists, if I'm not mistaken".

and subsequent events were to refute his belief, then the poor child will be justified in exclaiming:

"I *am* mistaken"!

Humbling moments like these are part of growing up. Note that this admission is formally identical to the Fool's paradox!

Evidently Kris Kringle, in his departure, left behind some fool's gold. How generous!

And were some sarcastic Grinch to snark:
"Santa Claus exists, and I am a liar".

and subsequent events were to confirm Santa's existence, then the Grinch would be justified in boasting:
"I am a liar"!

Recall that we can create Santa sentences by Grelling's method, by Quine's method, and by Russell's method:

Grelling's Santa:
Define the adjective "Santa-logical":
"A" is Santa-logical $=$ If "A" is A, then Santa exists.
Is "Santa-logical" Santa-logical?

"Santa-logical" is Santa-logical

    = If "Santa-logical" is Santa-logical, then Santa exists.

Quine's Santa is:

"Implies that Santa exists when quined" implies that Santa exists when quined.

Russell's "Santa Set for sentence G" is:

$$S_G = \{x | (x \in x) \text{ implies } G\}$$

Therefore: $x \in S_G = (x \in x) \Rightarrow G$

and therefore: $S_G \in S_G = (S_G \in S_G) \Rightarrow G$.

If there is no Santa Claus, then the above are all paradoxes.

Grinchian sentences and sets also exist, by the same methods; Grelling's, Quine's and Russell's, thus:

Define the adjective "Grinchian" thus:

"A" is Grinchian = Santa exists, and "A" is not A.
Therefore:

"Grinchian" is Grinchian

    = Santa exists, and "Grinchian" is not Grinchian.

Quine's Grinch is:
"Does not apply to itself, and Santa exists, when quined",

    does not apply to itself, and Santa exists, when quined.

The Russell's Grinch Set for sentence S is:

$$\text{RG}_S = \{x | S \wedge (x \notin x)\}.$$

Therefore: $x \in \text{RG}_S = S \wedge (x \notin x)$

and therefore: $\text{RG}_S \in \text{RG}_S = S \wedge (\text{RG}_S \notin \text{RG}_S)$.

If Santa exists, then the above are all paradoxes.

Above I told Barber-like stories about Santa sets. For instance, in another Spanish village, the barber takes weekends off; so he shaves all those, and only those, who shave themselves only on the weekend:

B shaves $M =$ If M shaves M, then it's the weekend.

Does the barber shave himself?

B shaves $B =$ If B shaves B, then it's the weekend.

When Monday rolls around, then (B shaves B) = paradox.

That village is watched by the watchmen, who watch all those, and only those, who watch themselves only when fortune smiles:

W watches $C =$ if C watches C, then fortune smiles.

Who watches the watchmen?

W watches $W =$ if W watches W, then fortune smiles.

If fortune ever frowns, then (W watches W) = paradox.

Recently that village saw the end of the Cold War, which ended all wars, and only those wars, which end themselves only if money talks:

CW ends $W =$ if W ends W, then money talks.

Did the Cold War end itself?

CW ends CW = if CW ends CW, then money talks.

Does money talk? If not, then (CW ends CW) = paradox.

That village's priest proclaimed this theological doctrine:

God blesses all those, and only those, who bless themselves only when there is peace:

G blesses S = If S blesses S, then there is peace.

Does God bless God?

G blesses G = If G blesses G, then there is peace.

Is there peace? If not, then (God blesses God) = paradox.

Recall Promenides the Cretan, who said;

"If I am honest, then *some* Cretan is honest".

How logical! But alas, this is equivalent to:

"If *all* Cretans are liars, then so am I".

Promenides sounds logical; but his statement still leaves open the possibility that every Cretan is a liar, including Promenides.

D.   Antistrephon

In the next few paragraphs I take the role of judge, and address the shades of Protagoras and Euathius.

Gentlemen, you have given me a dilemma. If Euathius is to win this case, then he must show that he has no obligation under the contract; but the contract says that he need not pay just if he loses the first case — which is this one. He wins if he loses and he loses if he wins; and the same goes for Protagoras.

If I find for Protagoras, then the judgement should go for Euathius; and if I find for Euathius, then the judgement should go for Protagoras. You wish me to declare sentence, but any sentence I declare will be an incorrect sentence, a false sentence. Therefore I declare:

This Sentence is False.

The Pseudomenon; a paradox, or half-truth. By the nature of this case, I can be only half-right; I can only half-satisfy you. In the interest of justice, I should take a position midway between yours, favoring neither side. Compromise is called for.

I therefore reformulate this case. I say that it is actually *two* cases being decided simultaneously. The first case is about the second half of the fee, to be awarded only if the second case is lost; and the second case is about the first half of the fee, to be awarded only if the first case is lost.

This is an artificial division of the original case; it would make no difference if the original case had an unequivocal solution. But here equivocation is necessary, and it works; for it is consistent for Protagoras to win the first case and Euathius to win the second. Upon recombining these results, we see that Protagoras can claim half the fee, having *won but lost*, and Euathius can keep the other half of the fee, having *lost but won*.

One final legal note: in this case, as is usual, Protagoras won if and only if Euathius lost:

i iff not j = t.

What is unusual about this case is that it's also true that Protagoras won if and only if Euathius *won*:

i iff j = t.

Stranger still: either Protagoras won and lost, or Euathius won and lost!

(i and not i) or (j and not j) = t.

E.   Infinity, Finitude and the Heap

In Chapter 1, I asked, what is the *parity* of infinity? Is it odd or even? That is, what is the limit of the sequence {t, f, t, f, ...}?

When we take the left limit of that sequence, we get I; right limit gets us J; paradoxes both, and fittingly so, since infinity = infinity + one, so infinity has the opposite of its own parity. Evidently it is in the nature of infinity to blur some details. This should come as no surprise; infinity is notoriously paradoxical.

Even approaching infinity yields paradox. In Chapter 1, I heaped together the paradoxes of The Heap, The First Boring Number, Berry's Paradox, and Finitude. They all had in common the vagueness of the boundary between the interesting and the uninteresting. Surely both types of integers exist; but where do they meet?

Assuming that we could find a number on the boundary (even though the search for such a number would be boringly long), then it would be interesting just as much as it is boring; which suggests an intermediate state.

So is "the first boring number" boring or not? Yes but no!

And what is "the smallest number that cannot be defined in less than 20 syllables"? In standard decimal nomenclature, that would be 127,777. (However, other naming schemes might name 127,777 in fewer than 20 syllables. As ever, uncertainty reigns.)

If you were to pile together 127,777 grains of sand, each 1 mm wide, then they will form a conical pile approximately 9.9 cm wide and half as tall; a small but respectable Heap. If you tried to move this Heap one grain at a time, laboring 5 seconds per grain, 8 hours per day, 5 days per week, then you will finish the job in approximately 4.5 weeks; a Heap of work.

"One plus the largest number defineable in less than 20 syllables" might be one plus "12 googol googol googol googol googol googol googol googol googol", or $1 + 1.2 * 10^{901}$. (This is if you allow the use of the word "googol", for 10^{100}. Other naming schemes yield even greater numbers.)

F. Game Paradoxes

Recall the definition of Hypergame: its initial position is the set of all "short" games — that is, all games that end in a finite number of moves. For one's first move in Hypergame, one may move to the initial position of any short game.

Is Hypergame short?

In the above I told the story of "the Mortal"; an unborn spirit who must now make this fatal choice; to choose some mortal form to incarnate as, and thus be be doomed to certain death.

The Mortal has a choice of dooms. Is the Mortal doomed?

The answer is that Hypergame is Finitude in disguise. Presumably the Mortal lives until the last interesting moment, then dies of boredom.

Recall my definition of the game Normalcy:

The move $N \rightarrow G$ is legal = the move $G \rightarrow G$ is not
    legal.

Is Normalcy normal? Let $G = N$:

The move $N \rightarrow N$ is legal = the move $N \rightarrow N$ is not
    legal.

This is a game-theory version of Russell's paradox. Normalcy is normal if and only if it is not. So is Normalcy normal? Yes but no.

In the above I told the story of the Rebel, who may become those, and only those, who do not remain themselves:

R may become $B = B$ may not become B.

Can the Rebel remain a Rebel? Yes but no.

Presumably Rebels play at Normalcy.

Chapter 8

The Continuum

A. Cantor's Paradox

Cantor's proof of the "non-denumerability" of the continuum relies on an anti-diagonal. For suppose we had a countable list of the real numbers:

$$R_1 = 0.D_{11}, D_{12}, D_{13}, D_{14} \cdots$$
$$R_2 = 0.D_{21}, D_{22}, D_{23}, D_{24} \cdots$$
$$R_3 = 0.D_{31}, D_{32}, D_{33}, D_{34} \cdots$$
$$\vdots$$

where D_{NM} is the Mth binary digit of the Nth number.

Then we define Cantor's anti-diagonal number:

$$C = 0. \sim D_{11}, \sim D_{22}, \sim D_{33}, \sim D_{44} \cdots$$

If $C = R_N$ for any N, then $D_{NX} = \sim D_{XX}$,

Therefore $D_{NN} = \sim D_{NN}$.; the pivot bit buzzes.

From this single buzzing bit Cantor deduces the existence of an infinity beyond infinity of real numbers! Was more ever made from less?

In diamond logic, the continuum is "semi-countable"; countable listings are possible, but they all contain paradox bits. The continuum is intermediate!

B.   Dedekind Splices

Recall the "paradox of the boundary":

What day is midnight?
Is noon A.M. or P.M.?
Is dawn day or night? Is dusk?
Which country owns the boundary?
Is zero positive or negative? (± 0?)

If a statement is true at point A and false at point B, then somewhere in-between lies a boundary. At any point on the boundary, is the statement true, or is it false?

To solve the paradox *of* the boundary, put a paradox *on* the boundary:

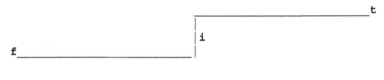

$$X \lesssim Y = (X < Y)\min(X \leq Y).$$

This is the "Dedekind splice" operator, equal to paradox at the boundary.

To make this a "continuous" function from R to diamond, we need to define a topology on diamond. Let the "open subsets" of diamond be the right-closed subsets:

$$\text{Open sets} = \{O : [(x \text{ in } O) \text{ and } (x < y)] \Rightarrow (y \text{ in } O)\}$$

$$= \{\{\}, \{j\}, \{t, j\}, \{f, j\}, \{t, f, j\}, \{i, t, f, j\}\}.$$

In this topology, all harmonic functions are continuous, the Dedekind splice is continuous, all values are near i, and none are near j.

The Dedekind splice is *anti-symmetric*, *transitive*, and *dense*:
For all x, y and z:
$(x \lesssim y) = \sim (y \lesssim x)$
if $(x \lesssim z)$ and $(z \lesssim y)$ then $(x \lesssim y)$
if $(x \lesssim y)$, then there exists a z such that $(x \lesssim z)$ and $(z \lesssim y)$.

If the sequence $\{x_n\}$ approaches the limit x from "both sides", as in an alternating series, then

$$(x \lesssim y) = \lim{}^- (x_n \lesssim y).$$

In general

$$((\lim x_n) \lesssim y) \preceq \lim{}^- (x_n \lesssim y).$$

The splice's anti-symmetry implies the paradox of the boundary:

$$(x \lesssim x) = \sim (x \lesssim x).$$

C. Null Quotients

The *null quotients* are the result of division by zero. There are two of them:

$1/0 =$ "infinity"; larger than any finite quantity.
$0/0 =$ "indefinity"; indistinguishable from any quantity.

Consider these algebraic equations:

$$x = 1/0,$$
$$0x = 1,$$
$$0 = 1.$$

Infinity leads us to an obvious absurdity. $1/0$ is inherently inconsistent; "over-determined".

Consider these algebraic equations:

$$x = 0/0,$$
$$0x = 0,$$
$$0 = 0.$$

Indefinity leads us to a vague tautology. $0/0$ is inherently uninformative; "underdetermined".

As noted in Chapter 2, this connects us to diamond logic; for we can identify i with one, and j with the other.

In terms of finitude paradoxes, perhaps we can say:

0/0 = the Heap, and 1/0 = Finitude

Consider the "sign" function:

$\text{sign}(x) = |x|/x.$

Its graph is:

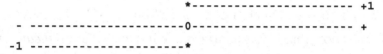

Note that $\text{sign}(0) = 0/0$; sign is *undefined* at zero. Note also the similarity of this graph to the Dedekind splice.

According to Gödel's Theorem, any arithmetical deductive system is either inconsistent or incomplete. Inconsistent is overdetermined, like 1/0; incomplete is underdetermined, like 0/0; thus arithmetic, though it avoids using null quotients, itself resembles a null quotient!

D.   Cantor's Number

Let us take a closer look at Cantor's anti-diagonal number; just what kind of quantity is it?

This number is so fraught with mathematical significance that it forces us to postulate a transfinite infinity of infinities; so surely it must, within itself, contain a transfinite amount of *information* about all those infinities. Otherwise the silly thing's just bluffing us!

We know that C has a buzz-bit. What does that bit *mean?* And how many such bits are there? Only one? A finite number? Infinitely many? Cofinitely many?

I can think of two interpretations of the buzz-bit:

$I = 1/2$. This is paradox as intermediate; a convergent compromise.

$I =$ "0 but 1". This is paradox as uncertainty; a divergent blur.

The 1/2 interpretation is like the "gap" interpretation; I as neither 0 nor 1. The blur interpretation is like the "glut" interpretation; I as both 0 and 1.

$I = 1/2$ means that it serves as a carry-and-borrow bit:

$.01 = .i0,$

$.10 = .i1.$

This implies higher-order carries:

$.0\ 1\ 1\ 1\ =\ .i\ i\ i\ 0,$

$.i\ i\ i\ 1\ =\ .1\ 0\ 0\ 0.$

In the infinite limit you get:

.0 1 1 1... = .i i i i... = .1 0 0 0....

This is the infinite bit-flip at each dyadic number $m/2^N$. It's also a version of Zeno's Paradox of Division, as given by the geometric sequence:

$$1 = 1/2 + 1/4 + 1/8 + 1/16 + \cdots .$$

C does not have I bits in cofinitely many digits; for an infinity of reals are all boolean; so the anti-diagonal has infinitely many boolean digits.

Perhaps C has an infinity of boolean bits, but an infinity of I bits too:

$$C = 0.01i11ii110i0i1100ii0\ldots.$$

In this interpretation, Cantor's paradox detects carry bits. C is uncleared!

The I = blur interpretation arises in the limit fixedpoints of Chapter 6. If an infinite iteration of a function on a real variable doesn't converge to a constant, then some digits of the number never converge. This endless motion between 0 and 1 is represented by I.

The number 0.i0i1 represents endless motion between four numbers:

0.0 0 0 1;   0.0 0 1 1;   0.1 0 0 1;   0.1 0 1 1.

Three I digits indicate eight numbers; N represent 2^N numbers; and a dispersed infinity of I digits represent a Cantor dust of real numbers.

So in this interpretation, C is a strange attractor; and its I bits tell where the other strange attractors are. That's more interesting than a carry bit, but even fractal chaos lacks the mystic glamour of "uncountable infinity".

So Cantor's number C can be a rough draft or a fractal; either way, what a comedown from transfinity! The silly thing *was* bluffing us!

Cantor's Theorem is hereby exposed as not only superfluous, but actually ridiculous. The continuum is countable; Cantor's Paradox detects carries or chaos. Therefore I propose a down-to-earth alternative to Cantor's tottering cardinal tower; a single countable infinity with paradox logic.

A slightly subtler logic yields an infinitely simpler model. This is known as elegance; sign of a correct theory.

If we insist that no real numbers on our list contain either carries or chaos, then Cantor's number still works as a boolean dyadic:

$$. 0\ 1\ 1\ 1 \ldots \ = \ . 1\ 0\ 0\ 0 \ldots$$

This entirely boolean C can fit anywhere on the list, for it's on the list with opposite *phase* from its appearance on the diagonal. This interpretation of C explains Cantor's Paradox — by *Zeno's* Paradox.

In this interpretation of Cantor's Number, the reals have all boolean bits, and they are countable; the anti-diagonal fits on the list due to dyadic bit-flip. Therefore the diagonal of the list is also dyadic; it ends in an infinite 1111 or 0000. The diagonal's limit is boolean.

I therefore propose this counter-Cantorian axiom:

Diagonal Limit. The diagonal of any complete list of boolean reals has a boolean limit.

E.   The Line within the Diamond

The "approximate comparison" operator $\gtrsim$ is ideal for embedding the continuum in diamond logic. Consider the following mapping from R (the continuum) to $\Diamond^\omega$ (the space of all infinite diamond-valued sequences):

$$E(x) = (x \gtrsim q_1, x \gtrsim q_2, x \gtrsim q_3, x \gtrsim q_4, \ldots),$$

where q_n is an enumeration of the rationals.

This function E sends R (the real number continuum) into $\Diamond^\omega$, the space of all infinite diamond vectors.

Its nth component, E_n, is comparison with the nth rational: $x \gtrsim q_n$.

Theorem. *This mapping E embeds R in $\Diamond^\omega$: that is, R's topology is carried intact into $\Diamond^\omega$, the space of diamond vectors.*

Proof. First, note that E is one-to-one; for if $x < y$, then some rational number q_n is between them; so

$$E_n(x) = f \quad \text{and} \quad E_n(y) = t.$$

Next note that E is continuous; for each of its components is continuous.

To complete proof of embedding, we need to prove this

Lemma. *The inverse of E is continuous.*

A function is continuous if the inverse image of an open set is an open set. The real line's topology is generated by the "half-lines":

$$(x, +\infty) = \{y : x < y\},$$
$$(-\infty, x) = \{y : y < x\}$$

so it suffices to prove that E sends each half-line to the intersection of an open set in $\diamond^\omega$ with the image $E(R)$.

$$E(x, +\infty) = \text{Union } [n \text{ such that } q_n > x]\{E(y) : E_n(y) = \text{t}\},$$
$$E(-\infty, x) = \text{Union } [n \text{ such that } q_n < x]\{E(y) : E_n(y) = \text{f}\}.$$

The first is a countable union of intersections of $E(R)$ with the open set $\{s : s_n = \text{t}\}$; the second is a countable union of intersections of $E(R)$ with the open set $\{s : s_n = \text{f}\}$. In either case, E sends a half-line to an intersection of $E(R)$ with an open set in $\diamond^\omega$.

Thus the lemma is proved: the inverse of E is continuous.

Therefore E is an embedding: 1-1 and bicontinuous.   QED

Theorem. *Any continuous function f from R to diamond "lifts" to a harmonic function f^* from $\diamond^\omega$ to diamond.*

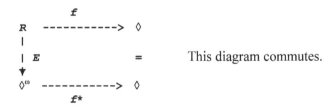

This diagram commutes.

Proof. Let $F(x)$ be a continuous function from R to diamond. The inverse image of an open set by a continuous function is an open set; so these are open sets:

$$F^-(t) = \{x \text{ in } R : F(x) = t\},$$
$$F^-(f) = \{x \text{ in } R : F(x) = f\}.$$

Call the first set A and the second set B. Being open, they are countable unions of open intervals:

$$A = \text{Union (all } N)(a_N, A_N),$$
$$B = \text{Union (all } N)(b_N, B_N),$$

where all the a's and b's are chosen from the rationals.

Approximate these sets by finite unions:

$$A_n(x) = (a_1 \lesssim x \lesssim A_1)$$
$$\vee\, (a_2 \lesssim x \lesssim A_2) \vee \cdots \vee (a_n \lesssim x \lesssim A_n),$$
$$B_n(x) = (b_1 \lesssim x \lesssim B_1)$$
$$\vee\, (b_2 \lesssim x \lesssim B_2) \vee \cdots \vee (b_n \lesssim x \lesssim B_n).$$

Then take left limits:

$$A(x) = \lim^- A_n(x)$$
$$= \lim^-((a_1 \lesssim x \lesssim A_1) \vee (a_2 \lesssim x \lesssim A_2)$$
$$\vee \cdots \vee (a_n \lesssim x \lesssim A_n)),$$
$$B(x) = \lim^- B_n(x)$$
$$= \lim^-((b_1 \lesssim x \lesssim B_1) \vee (b_2 \lesssim x \lesssim B_2)$$
$$\vee \cdots \vee (b_n \lesssim x \lesssim B_n)).$$

These are the characteristic functions for A and B; made strictly from the Dedekind splice and diamond logic.

Now define $F^*(x)$:

$$F^*(x) = A(x) \min {\sim} B(x)$$

$$= \lim{}^- ((a_1 \lesssim x \lesssim A_1) \vee (a_2 \lesssim x \lesssim A_2)$$

$$\vee \cdots \vee (a_n \lesssim x \lesssim A_n))$$

$$\min {\sim} (\lim{}^- ((b_1 \lesssim x \lesssim B_1) \vee (b_2 \lesssim x \lesssim B_2)$$

$$\vee \cdots \vee (b_n \lesssim x \lesssim B_n))).$$

This function is true if x is in the interior of A and the exterior of B: that is, $F(y) = \mathrm{t}$ and $F(y) \neq \mathrm{f}$, for any y near enough to x. This function equals false if x is in the exterior of A and the interior of B: that is, $F(y) = \mathrm{f}$ and $F(y) \neq \mathrm{t}$, for any y near enough to x. Finally, this function equals i at the boundary of the above two sets; that is, $F(y) = \mathrm{f}$, and $F(y') = \mathrm{t}$, for some y and y' in any neighborhood of x.

But F is a continuous function; so it equals t in the interior of A, f in the interior of B, and i at the boundary.

Therefore $F^*(x) = F(x)$.

Note that $F^*(x)$ is made only from E and diamond logic:

$$F(x) = F^*(x) = C(E(x)),$$

where C is a harmonic function.

Therefore $C(x)$ is a harmonic function which extends $F(x)$ (via the embedding E) to all $\diamond^\omega$.   QED

Theorem. *E is not only an embedding; it is a morphism: that is, functions from R to R "lift" to functions from $\Diamond^\omega$ to $\Diamond^\omega$;*

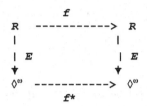

This diagram commutes.

Proof. If f is a function from R to R, then let f_n be the nth component of $f(x)$, via the embedding E:

$$f_n(x) = E_n(f(x)) = [f(x) \gtrsim q_n].$$

f is continuous; Dedekind splice is continuous; so f_n is continuous. Therefore, by the above theorem, f_n extends to a function from $\Diamond^\omega$ to diamond; therefore the function

$$f = (f_1, f_2, \ldots, f_n, \ldots)$$

extends to a continuous function from $\Diamond^\omega$ to $\Diamond^\omega$.   QED

Thus the real continuum embeds and extends into the space of diamond vectors. The continuum reduces to harmonic form.

F.   Zeno's Theorem

Every continuous function from the real line to itself extends to a harmonic function from diamond space to itself. But every harmonic function on diamond space has a fixedpoint.

Therefore we get:

Zeno's Theorem. *Any continuous function from the real line to itself has a fixedpoint in diamond space.*

I name this theorem after Zeno of Elea, famed for his paradoxes of motion. With the proof of this Theorem, we see that Zeno was right after all — in part. He claimed that no motion is possible: here we see that no motion is universal. Any continuous transformation of space has a fixedpoint; any chaotic dynamic has a paradoxical resolution.

G. Fuzzy Chaos

Consider "fuzzy logic", whose truth values are the real numbers between 0 and 1, where 0 means F and 1 means T. Fuzzy logic has these operators:

$$x \text{ and } y \;=\; \text{Minimum}(x, y),$$
$$x \text{ or } y \;=\; \text{Maximum}(x, y),$$
$$\text{not } x \;=\; 1 - x,$$

where minimum and maximum are relative to the usual ordering on the unit interval. As the previous section demonstrated, continuous real functions like these can be embedded into diamond space via the Dedekind splice. So:

$$x \text{ is "different from" } y = |x - y|,$$
$$x \text{ is "very true"} = x^2,$$
$$x \text{ is "nearly true"} = x^{1/2},$$
$$x \text{ is "extremely true"} = x^{16/5},$$
$$x \text{ is "slightly true"} = x^{5/16},$$
$$x \text{ is "at variance from" } y = (x - y)^2,$$
$$x \text{ "approximates" } y = 1 - (x - y)^2.$$

Note that "x is at variance from y" = "x is very different from y" and that "x approximates y" = "x is not very different from y".

Now allow fuzzy truth functions to self-refer dynamically. For instance the Liar paradox, in fuzzy logic, becomes this iteration:

$$P_{n+1} = 1 - P_n.$$

This has a constant solution $P_n = 0.5$, but also these wave solutions:

.1, .9, .1, .9, .1, .9, . . .

.7, .3, .7, .3, .7, .3, . . .

.4, .6, .4, .6, .4, .6, . . .

— and many others. Self-reference in fuzzy logic yields many different dynamical behaviors; neutral oscillations like this, convergence to a fixedpoint, convergence to a limit cycle, and "chaos".

For instance, consider the Boaster, who says, "I am very honest".

$$B_{n+1} = B_n^2.$$

This has an attracting fixedpoint at 0, and a repelling fixedpoint at 1.

By contrast, the Modest Truthteller says, "I am slightly honest".

$$M_{n+1} = M_n^{1/2}.$$

This has an attracting fixedpoint at 1, and a repelling fixedpoint at 0.

The Golden Liar says, "I am slightly untrue".

$$G_{n+1} = (1 - G_n)^{1/2}.$$

This has an attracting fixedpoint at 0.6180339888...; that is, $1/\phi$, or $\phi - 1$, where $\phi =$ the golden mean.

The Equivocal Liar says, "I am not very true".

$$E_{n+1} = 1 - E_n^2.$$

This has a repelling fixedpoint at $\varphi - 1$, and a limit cycle of 0,1,0,1,0,1,0,1,0,1,0,1,0,1,0,1, ...

Now consider the Chaotic Liar, who says,

"I do not differ from my negation".

$$C_{n+1} = 1 - |1 - C_n - C_n| = 1 - |1 - 2C_n|.$$

If you draw C_{n+1} versus C_n , you get a "tent" function, with a single peak. C_{n+2} versus C_n has two peaks, C_{n+3} versus C_n has four peaks, ... and C_{n+k} versus C_n has 2^{k-1} peaks.

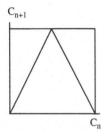 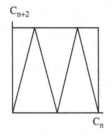 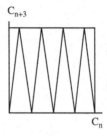

This indicates that C_n is a "chaotic" function, deterministic yet unpredictable, with sensitive dependence on initial conditions.

Now consider the Logistic Liar, who says,

"I am not very different from my negation".

That is, "I approximate my opposite".

$$L_{n+1} = 1 - (1 - L_n - L_n)^2.$$

So

$$L_{n+1} = 4L_n(1 - L_n).$$

This is none other than the logistic map, most studied of all chaotic dynamical systems. (A complex version of this map yields the Mandelbrot set.)

Now consider this Socratic dialog:

Socrates: "I approximate Plato's negation".

Plato: "I approximate Socrates".

$$S_{n+1} = 1 - (1 - P_n - S_n)^2,$$
$$P_{n+1} = 1 - (S_n - P_n)^2.$$

This system has a limit attractor in the form of a loop:

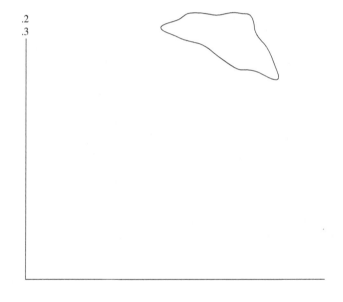

Now consider this dialog:

Socrates: "I am not even slightly different from Plato's opposite".

Plato: "I am not extremely different from Socrates".

$$S_{n+1} = 1 - |1 - P_n - S_n|^{5/16},$$

$$P_{n+1} = 1 - |S_n - P_n|^{16/5}.$$

It has a fractal attractor. Behold Zeno's Theorem gone gnarly:

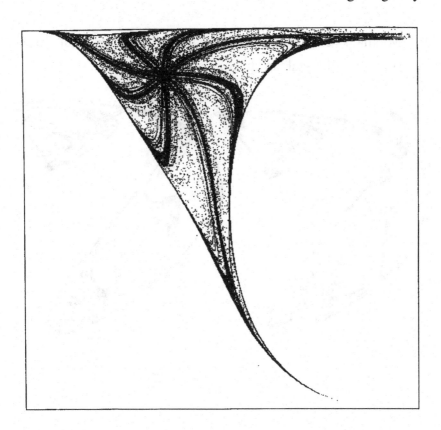

Here is a three-way conversation:

Moe: "I am not very different from Larry".

Larry: "I am not very different from Curly".

Curly: "I am very different from Moe".

$$M_{n+1} = 1 - (M_n - L_n)^2,$$

$$L_{n+1} = 1 - (L_n - K_n)^2,$$

$$K_{n+1} = (K_n - M_n)^2.$$

My thanks to Lou Kauffman for the following printout. To see the "Humpback Fractal" in stereo, just cross your eyes:

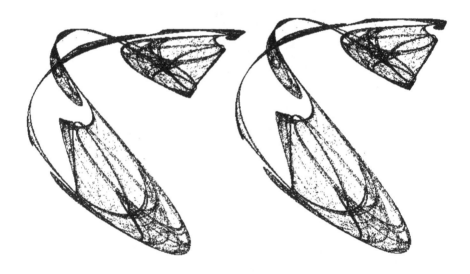

Chapter 9

Clique Theory

A. Cliques

Consider the Russell set; the set of all sets that do not contain themselves:

$R = \{x| \ x \notin x\}$.

In general: $(x \in R)$ = $(x \notin x)$

and therefore: $(R \in R)$ = $(R \notin R)$.

Therefore $R \in R$ is a paradox. In boolean logic, that means that R does not exist; which implies unspecified restrictions on set-building. But in diamond logic, paradox is no obstacle; $R \in R$ may equal I, or J, no problem. By accepting paradox, we defuse the Russell set.

In diamond logic, any property of sets is itself a set; even properties defined in terms of themselves. Diamond-valued sets are different enough from boolean sets that I give them another name; *cliques*.

Here's a close relative of Russell's set; the "Short-Circuit Clique":

$$S = \{x \mid S \notin S\}.$$

X is not a variable in the definition of S, so S is constant-valued:
For all x, $(x \in S) = (S \notin S) = (S \in S)$.

All cliques are paradox elements for S. What's more, S is defined in terms of itself; a no–no in boolean sets, but not a problem for cliques.

The empty clique is constant-valued, like S:

For all x, $\quad (x \in \{\}) = \text{F}$.

The empty clique is the falsehood predicate.
The world clique is the truth predicate:

For all x, $\quad (x \in W) = \text{T}$.

Here's the Liar clique:

$$L = \{x \mid x \notin L\}.$$

For all x, $\quad (x \in L) = (x \notin L)$.
The Liar clique contains all cliques not in the Liar clique.

Here's the Groucho clique:

$$G = \{x \mid G \notin x\}.$$

For all x, $\quad (x \in G) = (G \notin x)$.
The Groucho clique is the clique of all cliques that don't have the Groucho clique as a member. In particular, $(G \in G) = (G \notin G)$.

In general, any property of cliques defined as a harmonic function of the epsilon relation is also a clique. That is, if $H(a, b, c, \ldots)$ is a harmonic logic operator, then the following clique exists:

$$C_H = \{x \mid H(\ldots, x_M \in x_N, \ldots)\}.$$

That is, for all cliques x

$$(x \in C_H) = H(\ldots, x_M \in x_N, \ldots).$$

This is Weak Comprehension for the Weak Clique world; any property of cliques defined from harmonic self-reference on $\in$ defines a clique. Later we'll include the equality relation, to get Strong Comprehension for the Strong World.

For instance, here is the Upper Duck Clique:

$$UD = \{x \mid x \in UD \vee x \notin UD\}.$$

That is, for all cliques x, $(x \in UD) = (x \in UD) \vee (x \notin UD) = D(x \in UD)$.

The Lower Duck Clique is:

$$ld = \{x \mid x \in ld \wedge x \notin ld\} = \{x \mid d(x \in ld)\}.$$

We can write these definitions in set notation:

$$UD = UD \cup {\sim}UD,$$

$$ld = ld \cap {\sim}ld,$$

where $\sim$ stands in for the complementation sign. However, I prefer to use the same notation for cliques as I do for propositions, so

$$UD = UD \vee {\sim}UD,$$

$$ld = ld \wedge {\sim}ld.$$

The Truck clique is this system:

$L = \{x \mid x \notin L\}$,

$T = \{x \mid (x \in T) \vee (x \in L)\}$.

In clique notation, that is:

$L = \sim L$,

$T = T \vee L$.

$A = (x \in L)$ and $C = (x \in T)$ solve the Truck form:

$C = [[[A]_A C]]_C$.

Define the Santa Groucho clique as the clique of all cliques that contain the Santa Groucho clique only if Santa Claus exists:

$$SG \quad = \quad \{x \mid (SG \in x) \Rightarrow (\text{Santa exists})\},$$

$$(x \in SG) \quad = \quad ((SG \in x) \Rightarrow (\text{Santa exists})).$$

Define the Grinch Groucho clique as the clique of all cliques that do not contain the Grinch Groucho clique, and Santa Claus exists:

$$GG \quad = \quad \{x \mid (GG \notin x) \wedge (\text{Santa exists})\},$$

$$(x \in GG) \quad = \quad ((GG \notin x) \wedge (\text{Santa exists})).$$

Therefore:

$$(SG \in SG) \quad = \quad ((SG \in SG) \Rightarrow (\text{Santa exists})),$$

$$(GG \in GG) \quad = \quad ((GG \notin GG) \wedge (\text{Santa exists})).$$

If the first statement is boolean, then Santa Claus exists. If the second statement is boolean, then Santa Claus does not exist. Either way, one of these sentences is a Liar.

Given a clique C, its "description", $\text{Des}(C)$, is the clique of all its properties; that is, the clique of all cliques that contain C:

$$\text{Des}(C) \quad = \quad \{x \mid C \in x\},$$

$$(x \in \text{Des}(C)) \quad = \quad C \in x.$$

Recall the Groucho clique:

$$G = \{x \mid G \notin x\} = \ \sim\text{Des}(G).$$

The Groucho clique is the negation of its description. It is the clique of all the properties that it doesn't have.

Now consider this clique; the "Self-Description":

$$\text{SD} = \text{Des}(\text{SD}) = \{x \mid \text{SD} \in x\},$$

$$(x \in \text{SD}) \quad = \quad (\text{SD} \in x).$$

It contains its containers. In particular W, the world clique, is in SD, but not $\{\,\}$, the empty clique. In fact no well-founded clique is in SD; for if $x \in \text{SD}$, then $W \in W \in \text{SD} \in x$.

If x is in SD, and so is y, then so is their intersection; if x is in SD, and x is a subset of z, then z is in SD; and x is in SD to the exact extent that its negation $\sim x$ is not. Therefore SD is like an ultrafilter on the clique world.

But it isn't boolean:

$$(G \in \text{SD}) \quad = \quad (\text{SD} \in G) \quad = \quad (G \notin \text{SD}).$$

Note that if $x \in y$, then $y \in \text{Des}(x)$, so $\text{Des}(x) \in \text{Des}(y)$. Therefore the descriptions of the cliques have the same epsilon relations as the cliques. Des is an epsilon–morphism.

B. Clique Equality

The preceding was about the Weak Clique world, defined by harmonic self-reference on epsilon. But what of the equality property? Diamond logic's "if and only if" connective won't work: $(I$ iff $J) = $ T.

Also, shall we define equality *extensionally* — that is, in terms of clique *elements*–or *intensionally*–that is, in terms of clique *containers*?

$$(A =_E B) = (\text{For all } x)((x \in A) = (x \in B)),$$

$$(A =_I B) = (\text{For all } x)((A \in x) = (B \in x)).$$

Note that the property of containing a clique x is itself a property:

$$(x \in A) = (A \in \text{Des}(x))$$

so intensional equality implies extensional equality; it is a stricter notion. In boolean set theory, they are identical; sets with the same elements are the same; but not so in clique theory.

For consider the clique $\{\}'$:

$$\{\}' \quad = \quad \{x \,|\, \{\}' = \{\}\},$$

$$(x \in \{\}') \quad = \quad (\{\}' = \{\}),$$

$\{\}'$ is a constant-value clique, the same for all x. If $\{\}' = \{\}$, then every x is in $\{\}'$, unlike $\{\}$. Therefore $\{\}' \neq \{\}$; so $\{\}'$ has no elements, yet $\{\}'$ is unequal to $\{\}$, which also has no elements.

Therefore extensional equality fails for cliques; cliques can have the same elements yet still differ. In clique theory, equality is intensional; cliques are the same if they have the same containers.

The Strong Clique world is defined harmonically from both epsilon and equality, which is defined intensionally:

$$(x \in C_H) \quad = \quad H(\ldots, x_M \in x_N, \ldots, x_M = x_N, \ldots),$$

$$(A = B) \quad = \quad (\text{For all } x)((A \in x) = (B \in x)),$$

where each H is a harmonic logic operator.

For a given system of H's, can we define a clique world for it? Yes, we can! It requires an iteration longer than weak-clique iteration, but the iteration must stop, due to the phenomenon of "differentiation".

To construct a clique world, start with approximations for epsilon and equality. Let $\in_0$ be our initial epsilon, and $(x =_0 y)$ equal "true" for *any* x and y. This is the "initial unity"; default equality.

From $=_0$ and our initial $\in_0$, compute a new fixedpoint $\in_1$ by the limit method of Chapter 6:

$$(x \in_1 C_H) \quad = \quad \lim^- H^n(\lim^-(H^n(\in_0, =_0))) \quad = \quad H(\in_1, =_0).$$

Now define a new equality operator $=_1$:

$$(A =_1 B) \quad = \quad (\text{For all } x)((A \in_1 x) = (B \in_1 x)).$$

From $\in_1$ and $=_1$ define $\in_2$ and $=_2$ by similar means:

$$(x \in_2 C_H) \quad = \quad \lim^- H^n(\lim^-(H^n(\in_1, =_1))) \quad = \quad H(\in_2, =_1)$$

$$(A =_2 B) \quad = \quad (\text{For all } x)((A \in_2 x) = (B \in_2 x)).$$

Continue so on through the ordinals: $\in_3, =_3, \in_4, =_4, \ldots$; and at limit ordinals take limits.

If at any state of this iteration it so happens that $A \neq B$, despite the initial unity, then $A \neq B$ yet $B = B$; so A and B differ on the property $x = B$; so they will remain different at the next stage of the iteration. Therefore once cliques separate from each other, they remain separated. That is "differentiation".

Since terms can only become less equal, and never more equal, this iteration must stop sometime before every term becomes different from every other term. At some large countable ordinal α, equality stabilizes; then epsilon stabilizes soon after, and you get a stable system:

$$(x \in_\alpha C_H) \;=\; H(\ldots, x_M \in_\alpha x_N, \ldots, x_M =_\alpha x_N, \ldots),$$

$$(A =_\alpha B) \;=\; (\text{For all } x)((A \in_\alpha x) = (B \in_\alpha x)).$$

Cliques are defined simultaneously, in terms of each other, with default unity of equality and default paradox containment. So you do not "construct" a clique from the bottom up, as you do in Zermelo–Frankel set theory; instead you *invoke* a clique. You merely define it as a property of cliques, and its elements and properties grow organically with the rest of the clique world. The Zermelo–Frankel boolean set universe is like a tower; the clique world is like a web.

So for instance, we can define the Self-Singlet:

$$\text{SS} \;=\; \{\text{SS}\},$$

$$(x \in \text{SS}) \;=\; (x = \text{SS}).$$

If we define another clique the same way:

$$(x \in \text{ss}) = (x = \text{ss}),$$

then by symmetry of definition, ss and SS never separate from their initial unity, so they remain equal. In general, cliques given isomorphic definitions never differentiate, and so remain equal.

We'd like to define a doublet:

$A = \{A, B\}$,

$B = \{A, B\}$; yet A and B differ.

But the naive definition doesn't work:

$$(x \in A) \quad = \quad (x = A) \vee (x = B),$$
$$(x \in B) \quad = \quad (x = A) \vee (x = B)$$

because A and B never differentiate. To separate them you must add asymmetrical "differentiation terms":

$$(x \in A) \quad = \quad ((x = A) \vee (x = B)) \wedge (A \neq B),$$
$$(x \in B) \quad = \quad (x = A) \vee (x = B).$$

In the initial unity, the differentiation term $A \neq B$ makes A have no elements, unlike B; so they differ on the next turn, and thereafter.

Note that the differentiation term has no effect after differentiation; it is as if it is no longer there. Nonetheless the division it caused persists. A and B are unequal because they are unequal; a self-perpetuating separation that I call *arbitrary difference*.

By similar means we can make a triplet:

$$(x \in A) \quad = \quad ((x = A) \vee (x = B) \vee (x = C)) \wedge (A \neq B),$$
$$(x \in B) \quad = \quad ((x = A) \vee (x = B) \vee (x = C)) \wedge (B \neq C),$$
$$(x \in C) \quad = \quad ((x = A) \vee (x = B) \vee (x = C)).$$

So A, B and C all contain just A, B and C; yet they are unequal!

Of course we can make quadruplets and up.

C.   Clique Axioms

Clique theory has two main axioms; General Comprehension and Intensional Equality.

General Comprehension

Given any harmonic operator H, there is a clique C_H:

$$C_H \;=\; \{x \mid H(\ldots, x_M \in x_N, \ldots, x_M = x_N, \ldots)\}$$

such that for any clique x:

$$(x \in C_H) \;=\; H(\ldots, x_M \in x_N, \ldots, x_M = x_N, \ldots).$$

Intensional Equality

$$(x_A = x_B) \;=\; \text{(For all } x)((x_A \in x) = (x_B \in x)).$$

General Comprehension is clique theory's "wave-like" axiom. It posits epsilon as based on paradoxical self-reference. Intensional Equality is clique theory's "particle-like" axiom, where equality is as boolean and strict as possible. Thus clique theory combines equality's boolean sharpness with epsilon's paradoxical flexibility.

Clique theory fulfills some, but not all, of the axioms of Zermelo–Frankel set theory. It satisfies the limited comprehension, null set, pairs, unions, replacement, and infinity axioms; but not choice, foundation, or extensionality. And unlike the ZF sets, cliques include the world clique, negations, and iffs.

Limited Comprehension

Given a clique C and a clique property $P(x)$, there is a clique $C \wedge P$:

$$C \wedge P = \{x \mid x \in C \wedge P(x)\}.$$

Null Set

There is a null clique $\{\}$:

$$\{\} = \{x \mid \text{False}\}.$$

Pairs

Given cliques A and B, there is a pair clique $\{A, B\}$:

$$\{A, B\} = \{x \mid x = A \vee x = B\}.$$

Unions

Given a clique C, there is a union clique $\vee C$:

$$\vee C = \{x \mid \text{for some } z, x \in z \wedge z \in C\}.$$

In particular, there is the union of two sets:

$$A \vee B = \vee\{A, B\} = \{x \mid x \in A \vee x \in B\}.$$

Replacement

Given a clique C, and a function f on the cliques, there is a clique $\{f\}C$:

$$\{f\}C = \{x \mid \text{for some } z, x = f(z) \wedge z \in C\}.$$

Infinity

There is a clique Infinity:

$$\text{Infinity} = \{x \mid x = \{\} \vee (\text{for some } z)(z \in \text{Infinity} \wedge x = \{z\})\}.$$

The clique Infinity includes the cliques {}, {{}}, {{{}}}, {{{{}}}}, and so on. If you change the definition to end with "...$\wedge(x = z \vee \{z\})$", then you get the von Neumann ordinals;

$$\{\}, \ \{\{\}\}, \ \{\{\}, \{\{\}\}\}, \ \{\{\}, \{\{\}\}, \{\{\}, \{\{\}\}\}\}, \dots .$$

I think these are inefficient; they require 2^{N+1} brackets to define N.

The clique world does not obey the Foundation axiom, as the self-singleton set makes clear: $SS = \{SS\}$, so $\dots \in SS \in SS \in SS$.

Nor does it obey Extensionality, as the {} and {}′ cliques make clear.

The Choice axiom also fails. Given any function $f(x)$ on the cliques, define this *anti-choice clique*:

$$\mathrm{ACC}(f) = \{x \mid x \neq f(\mathrm{ACC}(f))\}$$

$\mathrm{ACC}(f)$ is the clique containing everything *except* its image under f!

In ZF set theory, the set universe is not a set itself. But in clique theory, the clique world is a clique, the constant truth function:

$$W = \{x \mid \text{True}\}.$$

Also, unlike ZF set theory, cliques include negations and iffs of cliques:

$$\sim C \ = \ \{x \mid x \notin C\},$$
$$C \text{ iff } D \ = \ \{x \mid x \in C \text{ iff } x \in D\}.$$

In general any harmonic function of cliques is a clique.

Define a *logomorphism* as a function from the clique terms to the clique terms that preserve definitions. For instance, consider these cliques:

$$(x \in A) \;=\; (x = B),$$

$$(x \in B) \;=\; (x = A).$$

The function $A \to B \to A$ is a logomorphism. But since their definitions are equivalent, A and B never differentiate, and they remain equal.

That is a special case of another clique theory axiom:

Logomorphic Equality

The only logomorphism is the identity.

D.   Graph Cliques

Because you can create a clique merely by invoking it, any finite graph can be made into a clique. For instance, take the "Goldfish Cliques":

$A = \{A\}$,

$B = \{A, C\}$,

$C = \{B, C\} = \{\{A, C\}, C\}$.

These cliques exist, from these definitions:

$(x \in A) \;\; = \;\; (x = A)$,

$(x \in B) \;\; = \;\; (x = A) \vee (x = C)$,

$(x \in C) \;\; = \;\; ((x = B) \vee (x = C)) \wedge (A \neq C)$.

The differentiation term is there to ensure separation.

Here's a cyclic clique:

$$A = \{B\},$$
$$B = \{C\},$$
$$C = \{A\},$$

$(x \in A) \;\; = \;\; (x = B)$,

$(x \in B) \;\; = \;\; (x = C) \wedge (B \neq C)$,

$(x \in C) \;\; = \;\; ((x = A) \wedge (C \neq A))$.

You can of course define longer cycles.

Infinite graphs also define graph cliques, given clear definitions. For instance we can define the order type of Z in cliques:

$$\ldots \in C_{-2} \in C_{-1} \in C_0 \in C_1 \in C_2 \in \ldots$$

$\ldots$

$$
\begin{aligned}
(x \in C_{-2}) \;=\;\; &(\text{for some } N)(N \le -2 \wedge x = C_N) \\
&\wedge (C_{-1} \ne C_{-2}), \\
(x \in C_{-1}) \;=\;\; &(\text{for some } N)(N \le -1 \wedge x = C_N) \wedge (C_0 \ne C_{-1}), \\
(x \in C_0) \;=\;\; &(\text{for some } N)(N \le 0 \wedge x = C_N), \\
(x \in C_1) \;=\;\; &(\text{for some } N)(N \le 1 \wedge x = C_N) \wedge (C_0 \ne C_1), \\
(x \in C_2) \;=\;\; &(\text{for some } N)(N \le 2 \wedge x = C_N) \wedge (C_1 \ne C_2).
\end{aligned}
$$

$\vdots$

There are even cliques with the order type of R; *continuous cliques*.

$$(x \in CC_r) \;=\; (\text{for some real } s)((x = CC_s) \wedge (s \le r)) \wedge \mathrm{DT},$$

where DT is a differentiation term, if r is rational.

There is a CC_r for each real number r, and

$$(CC_r \in CC_s) \;=\; (r \le s).$$

E. Clique Circuits

Starting from a well-founded set, no infinite descending element chains exist. This suggests a game; *Clique Nim*.

In Clique Nim, the first player chooses an element of C; call that C_1. The second player chooses an element of C_1; call that C_2. And so on. If the chain stops, then the player who cannot choose further loses.

Let U be the clique of all cliques for which the first player has a winning strategy; let V be the clique of all cliques for which the second player has a winning strategy. Then $U = \sim V$. For instance, $\{\}$ is in V but not U; $\{\{\}\}$ is in U but not V. $\{\{\}, \{\{\}\}\}$ is in U, because the first player can choose $\{\}$.

A clique x is in U — that is, the first player wins — if the second player loses for some one of x's elements — that is, some of x's elements are not in U:

$$U = \{x \mid (\text{For some } z)(z \in x \wedge z \notin U)\}.$$

A clique x is in V — that is, the first player loses — if the second player loses for none of x's elements — that is, all of x's elements are not in V:

$$V = \{x \mid (\text{For all } z)(z \in x \Rightarrow z \notin V)\}.$$

Therefore:

$$\{a, b, c, d, \ldots\} \in U$$
$$= (a \in U) \text{ nand } (b \in U) \text{ nand } (c \in U) \text{ nand} \ldots,$$
$$\{a, b, c, d, \ldots\} \in V$$
$$= (a \in V) \text{ nor } (b \in V) \text{ nor } (c \in V) \text{ nor} \ldots.$$

You can tell if a clique is in U or V by replacing the set brackets {} with boundary-logic brackets [ ]. If it reduces to null, then it's in U; and if it reduces to [ ], then it's in V. U reads {} as a nand gate, V reads {} as a nor gate.

For instance {} corresponds to [], so it's in V. The set {{}, {{}}} corresponds to the form [[][[]]] = [[]] = null; so it's in U.

If the cliques $a, b, c, d, \ldots$ form a graph clique as defined in the preceding section, then the clique inclusion statements $a \in U, b \in U$, $c \in U, d \in U, \ldots$ solve the equations for a logic fixedpoint; that is, they make a diamond circuit. So do the statements $a \in V, b \in V$, $c \in V, d \in V, \ldots$. These are the *clique circuits*.

For instance, the Self-Singleton SS = {SS} yields buzzer circuits:

$$(\text{SS} \in U) \quad = \quad \sim\text{SS} \in U,$$
$$(\text{SS} \in V) \quad = \quad \sim\text{SS} \in V.$$

Or take the Goldfish clique as defined in the previous section:

$A = \{A\},$

$B = \{A, C\},$

$C = \{B, C\} = \{\{A, C\}, C\}.$

This yields dual versions of the "goldfish" circuit:

$$(A \in U) \quad = \quad \sim(A \in U),$$
$$(C \in U) \quad = \quad (C \in U) \Rightarrow (A \in U \wedge C \in U),$$
$$(A \in V) \quad = \quad \sim(A \in V),$$
$$(C \in V) \quad = \quad (C \notin V) \wedge (A \in V \vee C \in V).$$

Now consider the *Brownian clique:*

$A = \{Z, D\}$,

$B = \{A, G\}$,

$C = \{D, B\}$,

$D = \{C, F\}$,

$E = \{C, Z\}$,

$F = \{E, H\}$,

$G = \{A, H\}$,

$H = \{E, G\}$.

These inclusions are somewhat symmetrical; the definition would require differentiation terms. Take $\in U$ or $\in V$ of these cliques, and you get the *First Brownian Modulator Circuit*, to be discussed below.

Part Two

The Second Paradox

Chapter 10

Orthogonal Logics

A.   Analytic Functions

If a function is harmonic, then it preserves order; but not all functions preserve lattice order, so not all function are harmonic. Consider these:

x	i	t	f	j
-(x) = termwise negation	j	f	t	i
*(x) = dual of x = ~ - x	j	t	f	i
l(x) = left side = x/*x	t	t	f	f
r(x) = right side = *x/x	t	f	t	f
L(x) = turn left = *x/~(x)	f	i	j	t
R(x) = turn right = ~(x)/*x	t	j	i	f

"Star" (*) obeys these laws:

$$^*M(x, y, z) = M(^*x, ^*y, ^*z),$$

$$^*(x \wedge y) = {}^*x \wedge {}^*y; \quad {}^*(x \vee y) = {}^*x \vee {}^*y,$$

$$^*(x \min y) = {}^*x \max {}^*y; \quad {}^*(x \max y) = {}^*x \min {}^*y,$$

$$^*(\sim(x)) = \sim ({}^*(x)) = -x;$$

$$^{**}x = x.$$

171

The left and right side operators obey these equations:

$$l(M(x, y, z)) = M(lx, ly, lz); \quad r(M(x, y, z)) = M(rx, ry, rz);$$
$$l(x \wedge y) = lx \wedge ly; \qquad\qquad l(x \vee y) = lx \vee ly;$$
$$l(x \min y) = lx \vee ly; \qquad\qquad l(x \max y) = lx \wedge ly;$$
$$r(x \wedge y) = rx \wedge ry; \qquad\qquad r(x \vee y) = rx \vee ry;$$
$$r(x \min y) = rx \wedge ry; \qquad\qquad r(x \max y) = rx \vee ry;$$
$$l(l(x)) = r(l(x)) = l(x); \qquad\qquad l(r(x)) = r(r(x)) = r(x);$$
$$l(\sim x) = rx; \; r(\sim x) = lx;$$
$$*(lx) = lx; \; *(rx) = rx;$$
$$l(*x) = rx; \; r(*x) = lx.$$

The rotation operators R and L obey these laws:

$$L(M(x, y, z)) = M(Lx, Ly, Lz); \quad R(M(x, y, z)) = M(Rx, Ry, Rz);$$
$$L(x \wedge y) = Lx \max Ly; \qquad\qquad L(x \vee y) = Lx \min Ly;$$
$$L(x \min y) = Lx \wedge Ly; \qquad\qquad L(x \max y) = Lx \vee Ly;$$
$$R(x \wedge y) = Rx \min Ry; \qquad\qquad R(x \vee y) = Rx \max Ry;$$
$$R(x \min y) = Rx \vee Ry; \qquad\qquad R(x \max y) = Rx \wedge Ry;$$
$$LLx = RRx = -x; \; LLLx = Rx; \quad RRRx = Lx; \; LRx = RLx = x;$$
$$L(\sim x) = \sim(Rx) = (-x/x); \qquad R(\sim x) = \sim(Lx) = (x/-x);$$
$$L(*x) = *Rx = (x/-x); \qquad\qquad R(*x) = *Lx = (-x/x).$$

B. Function Types

"Minus" reveals diamond's underlying boolean structure. Call a function F "analytic" if you can define minus by using it and the harmonic functions:

$$-(x) = G(F(H(x))), \quad \text{for some harmonic } G \text{ and } H.$$

Minus does not preserve lattice order; so no analytic function does either. The converse is also true:

If F does not preserve order, then F is analytic.

Or, to be more specific:

If $F(a) \not\preceq F(b)$ for some $a \preceq b$,

then at least one of these two functions equals $-(x)$:

$$F(a \max(b \min \lambda(x)))/{\sim} F(a \max(b \min \rho(x))),$$

$$\sim F(a \max(b \min \lambda(x)))/F(a \max(b \min \rho(x))).$$

Proof. Here are the truth tables for those two functions:

	i	t	f	j
F1	F(a)/~F(a)	F(a)/~F(b)	F(b)/~F(a)	F(b)/~F(b)
F2	~F(a)/F(a)	~F(a)/F(b)	~F(b)/ F(a)	~F(b)/F(b)

There are seven possible ways to have $F(a) \not\preceq F(b)$:

```
(F(a),F(b))  =

(j,t),   (j,i),   (j,f),   (t,i),   (t,f),   (f,i),  (f,t)


F1,F2 ; F1,F2 ; F1,F2 ; F1,F2 ; F1,F2 ; F1,F2;  F1,F2
```

i	j, j	j, j	j, j	i, j	i, j	j, i	j,i
t	f, j	f, f	f, f	i, f	t, f	f, i	f,t
f	t, j	t, t	t, t	i, t	f, t	t, i	t,f
j	i, j	i, i	i, i	i, i	j, i	i, i	i,j
	*	* *	* *	*	*	*	*

In each case at least one of $F1$ and $F2$ equals $-(x)$.   QED

If F is harmonic, then it preserves order. The converse is also true:

If F preserves order, then F is harmonic.

Proof. Note that if F preserves order, then

$$F(\mathrm{I}) \preceq F(\mathrm{T}) \preceq F(\mathrm{J}),$$

$$F(\mathrm{I}) \preceq F(\mathrm{F}) \preceq F(\mathrm{J}).$$

Therefore:

$$F(\text{I}) \min F(\text{T}) \min F(\text{F}) = F(\text{I}),$$

$$F(\text{J}) \max F(\text{T}) \max F(\text{F}) = F(\text{J}).$$

And also:

$$F(\text{T}) \vee (F(\text{I}) \wedge F(\text{J}))$$

$$= (F(\text{T}) \min(F(\text{I}) \max F(\text{J})))/(F(\text{T}) \max(F(\text{I}) \min F(\text{J}))$$

$$= (F(\text{T}) \min F(\text{J}))/(F(\text{T}) \max F(\text{I}))$$

$$= F(\text{T})/F(\text{T}) = F(\text{T}).$$

And similarly:

$$F(\text{F}) \vee (F(\text{I}) \wedge F(\text{J})) = F(\text{F}).$$

Now use the Differential Normal Form:

$$F^*(x) = (F(\text{T}) \wedge x) \vee (F(\text{F}) \wedge \sim x) \vee \text{M}(F(\text{I}), dx, F(\text{J})),$$

$$F^*(\text{T}) = (F(\text{T}) \wedge \text{T}) \vee (F(\text{F}) \wedge \text{F}) \vee \text{M}(F(\text{I}), F, F(\text{J}))$$

$$= F(\text{T}) \vee (F(\text{I}) \wedge F(\text{J})) = F(\text{T}),$$

$$F^*(\text{F}) = (F(\text{T}) \wedge F) \vee (F(\text{F}) \wedge \text{T}) \vee \text{M}(F(\text{I}), F, F(\text{J}))$$

$$= F(\text{F}) \vee (F(\text{I}) \wedge F(\text{J})) = F(\text{F}),$$

$$F^*(\text{I}) = (F(\text{T}) \wedge \text{I}) \vee (F(\text{F}) \wedge \text{I}) \vee \text{M}(F(\text{I}), \text{I}, F(\text{J}))$$

$$= (F(\text{T}) \wedge \text{I}) \vee (F(\text{F}) \wedge \text{I}) \vee (F(\text{I}) \min F(\text{J})$$

$$= (F(\text{T}) \wedge \text{I}) \vee (F(\text{F}) \wedge \text{I}) \vee F(\text{I})$$

$$= (F(\text{T}) \wedge \text{I}) \vee (F(\text{F}) \wedge \text{I}) \vee (F(\text{I}) \wedge \text{I})$$

$$\vee (F(\text{T}) \wedge F(\text{F}) \wedge F(\text{I})) \vee (F(\text{I}))$$

$$= (F(\text{T}) \min F(\text{F}) \min F(\text{I})) \vee (F(\text{I}))$$

$$= F(\text{I}) \vee (F(\text{I})) = F(\text{I}).$$

And similarly, $F^*(\text{J}) = F(\text{J})$.

Therefore $F = F^*$, so F is harmonic. QED

Therefore we have these two equivalences:

F is harmonic if and only if F preserves phase order.
F is analytic   if and only if F does not preserve phase order.

There are two kinds of functions on diamond; analytic or else harmonic.

C.   Dihedral Conjugation

Given a permutation P, a function F, and a relation R, we can define the function $P[F]$ and the relation $P[R]$ by:

$$P[F](x) = P(F(P^{-1}(x))),$$

$$xP[R]y \text{ iff } (P^{-1}(x))R(P^{-1}(y)).$$

These are F and R conjugated by P.

Conjugation Theorems.

$$
\begin{aligned}
P(F(x, y)) &= P[F](P(x), P(y)), \\
xRy &\text{ iff } P(x)P[R]P(y), \\
P[=] &= (=), \\
P[Q[F]] &= (P \circ Q)[F], \\
P[F] \circ (P[G]) &= P[F \circ G].
\end{aligned}
$$

Proof.

$$P[F](P(x), P(y)) = P(F(P^{-1}(P(x)), P^{-1}(P(y))))$$

$$= P(F(x, y)). \qquad \text{QED}$$

$$P(x)P[R]P(y) \text{ iff } P^{-1}(P(x))RP^{-1}(P(y)) \text{ iff } xRy. \qquad \text{QED}$$
$$xP[=]y \text{ iff } P^{-1}(x) = P^{-1}(y) \text{ iff } x = y. \qquad \text{QED}$$

$$P[Q[F]](x) = P(Q[F](P^{-1}(x))) = P(Q(F(Q^{-1}(P^{-1}(x)))))$$

$$= (P \circ Q) \circ F \circ (P \circ Q)^{-1}(x) = (P \circ Q)[F](x).$$

$$\text{QED}$$

$$P[F] \circ (P[G])(x) = P(F(P^{-1}(P(G(P^{-1}(x))))))$$

$$= P(F(G(P^{-1}(x))))$$

$$= P[F \circ G](x). \qquad \text{QED}$$

Whatever equational identities the functions F and G may have, the functions $P[F]$ and $P[G]$ also have. Thus the conjugate of a DeMorgan algebra is a DeMorgan algebra, the conjugate of a field is a field, etc. Conjugation transports identities.

Now let the dihedral group D operate on the diamond. It has four reflections and four rotations:

$(tf) = $ "not"; $(ij) = $ "*"; $(ti)(jf) = $ "o/ $-$ "; $(tj)(if) = $ " $-$ /o";
identity $= $ "o"; $(tif j) = $ "L"; $(tf)(ij) = $ " $-$ "; $(tjfi) = $ "R".

b

a*b	o	R	-	L	not	o/-	*	-/o
o	o	R	-	L	not	o/-	*	-/o
R	R	-	L	o	o/-	*	-/o	not
-	-	L	o	R	*	-/o	not	o/-
L	L	o	R	-	-/o	not	o/-	*
not	not	-/o	*	o/-	o	R	-	L
o/-	-/o	*	o/-	not	R	-	L	o
*	*	o/-	not	-/o	-	L	o	R
-/o	o/-	not	-/o	*	L	o	R	-

(with a labelling the rows.)

If we identify the two-dimensional real vectors with a "diamond vector"; that is, linear combinations of diamond values:

$$\binom{r}{s} = rt + si$$

then we can identify this group as 2×2 matrices;

$$o = \begin{pmatrix} 1 & 0 \\ 0 & 1 \end{pmatrix} \quad ; \quad \text{not} = \begin{pmatrix} -1 & 0 \\ 0 & 1 \end{pmatrix}$$

$$\mathrm{R} = \begin{pmatrix} 0 & 1 \\ -1 & 0 \end{pmatrix} \; ; \; \mathrm{o}/- = \begin{pmatrix} 0 & 1 \\ 1 & 0 \end{pmatrix}$$

$$- = \begin{pmatrix} -1 & 0 \\ 0 & -1 \end{pmatrix} \; ; \; * = \begin{pmatrix} 1 & 0 \\ 0 & -1 \end{pmatrix}$$

$$\mathrm{L} = \begin{pmatrix} 0 & -1 \\ 1 & 0 \end{pmatrix} \; ; \; -/\mathrm{o} = \begin{pmatrix} 0 & -1 \\ -1 & 0 \end{pmatrix}$$

Modulo $-$, these are equivalent to the generators of M(2, 2), the 2×2 matrices.

D permutes functions and relations as well as elements, by conjugation. For *P* in the dihedral group, this applies:

$$P[\mathrm{M}] = \mathrm{M}$$

That is:

$$P(\mathrm{M}(x, y, z)) = \mathrm{M}(P(x), P(y), P(z)).$$

Thus the above group table also defines the group's conjugation action on the logic operators:

		b					
	a[*b*]	**and**	**or**	**min**	**max**	**not**	*****
a	**o**	and	or	min	max	not	*
	R	min	max	or	and	*	not
	-	or	and	max	min	not	*
	L	max	min	and	or	*	not
	not	or	and	min	max	not	*
	o/-	max	min	or	and	*	not
	*****	and	or	max	min	not	*
	-/o	min	max	and	or	*	not

Note that all four positive operators distribute over each other; symmetrically.

D.   Star Logic

Most of the best properties of diamond are shared by 3-valued logic, a simpler sublogic; T, I and F. (Or equivalently; T, J and F.) 3-logic is a DeMorgan logic; it has enough semi-lattice to prove self-reference; it too solves the paradoxes of self-reference and continuity; it too embeds the continuum, and it too supports a clique theory. So what is the second paradox value doing?

The answer is that diamond-logic has inner symmetries unavailable to 3-logic; and that in fact there is a "paraharmonic" logic complementary to the harmonic logic.

Recall that $-(a/b) = (-a)/(-b)$; "termwise" negation.

Minus gets down to Diamond's boolean innards.

Let $* = \sim -$; that is, $*(a/b) = (*b)/(*a)$.

Star reverses order. It exchanges i and j, leaving t and f fixed. Thus Star looks just like Not, at "right angles"; it is "sideways negation".

In diamond logic star is flip:

$$*(a/b) = (*b)/(*a).$$

In dynamic implementation star equals delay:

$$(*a)(n) = a(n - 1).$$

In dual-rail circuits star equals swap wires.

Star, not, identity and — form a Klein group:

```
                      b

        ab  |  o   ~   *   -
        --------------------------
         o  |  o   ~   *   -
   a     ~  |  ~   o   -   *        Each element is
         *  |  *   -   o   ~        its own inverse
         -  |  -   *   ~   o
```

Let "star logic" be a logic made from star, majority, and the four values, just as diamond logic is made from not, majority, and the four values.

Star logic is isomorphic to diamond logic via rotation; therefore all results from the preceding chapters apply:

Star logic is a complete De Morgan algebra.

It proves the self-reference theorem.

It has limit operators.

The continuum embeds via a morphism.

Zeno's theorem.

Clique theory.

When we combine star logic and diamond logic, then we get minus, an operator with no fixedpoint. In a sense, then, star logic is "perpendicular" to diamond logic; similar to it, but at right angles. Therefore I call star logic "paraharmonic"; it resembles harmonic logic but is incompatible. Diamond logic is two-dimensional; it has room for two separate dimensions of thought within it. Negation and star are "perpendicular" logics; they work at cross-purposes.

E. Harmonic Projection

When we rotationally conjugate the "left side" and "right side" operators;

$$l(x) = x/^*x, \quad r(x) = {}^*x/x$$

we get two "harmonic projection" operators;

$$\lambda(x) = x/\sim(x); \quad \rho(x) = \sim(x)/x.$$

These send the diamond to i and j, just as l and r send the diamond to t and f. Unlike l and r, λ and ρ are harmonic, i.e. they preserve phase order. Being conjugate to l and r, they obey similar equations:

$\lambda(i) = i; \ \lambda(t) = i;$ $\qquad \lambda(f) = j; \ \lambda(j) = j;$

$\rho(i) = i; \ \rho(t) = j;$ $\qquad \rho(f) = i; \ \rho(j) = j;$

$\lambda(M(x, y, z)) = M(\lambda x, \lambda y, \lambda z);$ $\quad \rho(M(x, y, z)) = M(\rho x, \rho y, \rho z);$

$\lambda(x \wedge y) = \lambda x \max \lambda y;$ $\qquad \lambda(x \vee y) = \lambda x \min \lambda y;$

$\lambda(x \min y) = \lambda x \min \lambda y;$ $\qquad \lambda(x \max y) = \lambda x \max \lambda y;$

$\rho(x \wedge y) = \rho x \min \rho y;$ $\qquad \rho(x \vee y) = \rho x \max \rho y;$

$\rho(x \min y) = \rho x \min \rho y;$ $\qquad \rho(x \max y) = \rho x \max \rho y;$

$\lambda(\sim x) = \rho x; \ \sim(\lambda x) = \lambda x;$ $\quad \rho(\sim x) = \lambda x; \ \sim(\rho x) = \rho x;$

$\lambda(^*x) = {}^*\rho x;$ $\qquad\qquad \rho(^*x) = {}^*\lambda x.$

Harmonic logic can side-analyze star logic; and vice versa.

F.   Diamond Types?

The very concept of a diamond type seems ironic; for the whole point of diamond is to create typeless fixedpoints. In complete self-reference, there is but one type, and it refers to itself, harmonically or paraharmonically.

If we allow reference by analytic functions as well, then we need type theory; for analytic functions unearth diamond's boolean substrate. Stranger still, harmonic and paraharmonic functions are analytic *relative to each other*; that is, paraharmonic functions are analytic to harmonic systems, and vice versa. This is the "perpendicularity" of the two logics. Given two perpendicular systems, one must be of lower type than the other.

So in a diamond type theory, harmonic systems refer to paraharmonic systems, which refer to harmonic systems, down and down, each system orthogonal to the one below it. This is descent plus rotation; a helix!

Chapter 11

Interferometry

Quadrature
Diffraction
Buzzers and Toggles
Analytic Diffraction
Diffracting "Two Ducks in a Box"

A. Quadrature

a.k.a. general quadratic distribution

Theorem. *Junction Diamond*

For any diamond values x and y, we have:

$$
\begin{array}{ccc}
 & b & \\
\leq & \leq & \{a, c\} = \{x, y\} \quad and \quad \{b, d\} = \{x/y, y/x\} \\
a \quad \leq \quad c & & OR \\
\leq & \leq & \{a, c\} = \{x/y, y/x\} \quad and \quad \{b, d\} = \{x, y\}. \\
 & d &
\end{array}
$$

where:

Proof is by cases.

The "junction diamond" $\{x, y, x/y, y/x\}$ contains its own minimum, maximum, conjunction, and disjunction.

If f is a harmonic function, then it preserves order; hence we get a similar order diamond for $\{f(x), f(y), f(x/y), f(y/x)\}$. This implies this theorem:

Quadrature (or; **General Quadratic Distribution**):

For any harmonic function $f(x)$;

$$f(x \max y) = f(x) \max f(y) \max f(x/y) \max f(y/x);$$

$$f(x \min y) = f(x) \min f(y) \min f(x/y) \min f(y/x).$$

What is more, $f(x \max y)$ *equals* one of the four terms on the right; and similarly for $f(x \min y)$.

Corollary. *For any harmonic f;*

$$f(x \max y) \succeq f(x) \max f(y);$$

$$f(x \min y) \preceq f(x) \min f(y).$$

N.B. $\{f(x), f(y), f(x/y), f(y/x)\}$ may have other $\preceq$ and $=$ relations in addition to the "junction diamond". For instance:

$$di = i < dt = f = df < dj.$$

The derivative collapses the diamond to a linear order.

$$\lambda(i) = \lambda(t) = i < j = \lambda(f) = \lambda(j).$$

Harmonic projection collapses the diamond to two values.

Theorem. *Quadrature in k terms*:

$$f(x_1 \min x_2 \min \ldots \min x_k) = \text{Min}[i, j \le n](f(x_i/x_j));$$

$$f(x_1 \max x_2 \max \ldots \max x_k) = \text{Max}[i, j \le n](f(x_i/x_j)).$$

This has k^2 terms.

Theorem. *Quadrature in k terms, and n dimensions*:

If $x_1, x_2, x_3, \ldots, x_k$ are all n-dimensional vectors;

$$x_i = (x_{i,1}, x_{i,2}, \ldots, x_{i,n})$$

and $F(x)$ is from $\diamond^n$ to $\diamond$, then;

$$F(\mathrm{Min}[1 \leq i \leq k](x_i)) = \mathrm{Min}[1 \leq \text{all } i, j \leq n]$$
$$\times (F(x_{i1,1}/x_{j1,1}, \ldots, x_{in,n}/x_{jn,n}))$$
$$F(\mathrm{Max}[1 \leq i \leq n](x_i)) = \mathrm{Max}[1 \leq \text{all } i, j \leq n]$$
$$\times (F(x_{i1,1}/x_{j1,1}, \ldots, x_{in,n}/x_{jn,n})).$$

This has k^{2n} terms.

Theorem. *Quadratic continuity*:

For all harmonic functions $f(x)$

$$f(\lim{}^- x_n) = \lim{}^- f(x_i/x_j);$$
$$f(\lim{}^+ x_n) = \lim{}^+ f(x_i/x_j),$$

where the second limits are of the form:

$$\mathrm{Max}[\text{all } N]\mathrm{Min}[\text{all } i, j > N];$$

$$\mathrm{Min}[\text{all } N]\mathrm{Max}[\text{all } i, j > N],$$

i.e. *"in the limit of large i and j"*.

Proof is via "cofinal-range" definition of the limit operators. Recall that the $\lim^-$ of a sequence is the minimum of the "cofinal range", i.e. the set of all values that occur infinitely many times:

$$\lim{}^-\{x_n\} = \mathrm{Min} \text{ cofinal}\{x_n\};$$
$$\lim{}^+\{x_n\} = \mathrm{Max} \text{ cofinal}\{x_n\};$$

where $\mathrm{cofinal}\{x_n\} = \{Y : x_n = Y \text{ for infinitely many } n\}$.

Therefore

$$f(\lim^-\{x_n\}) = f(\text{Min cofinal}\{x_n\})$$

$$= \text{Min } f((\text{cofinal}\{x_i\})/(\text{cofinal}\{x_j\}))$$

$$= \text{Min } f((\text{cofinal } \{x_i/x_j\}))$$

$$= \text{Min cofinal } \{f(x_i/x_j)\}$$

$$= \lim^- f(x_i/x_j).$$

Similarly,

$$f(\lim^+\{x_n\}) = \lim^+ f(x_i/x_j).$$

Theorem. *Phased distribution over the positives*

$$f(a \wedge b) = (f(a) \min f(a/b)) \quad \max \quad (f(b) \min f(b/a))$$

$$= (f(a) \max f(b/a)) \quad \min \quad (f(b) \max f(a/b));$$

$$f(a \vee b) = (f(a) \min f(b/a)) \quad \max \quad (f(b) \min f(a/b))$$

$$= (f(a) \max f(a/b)) \quad \min \quad (f(b) \max f(b/a)).$$

These are "permeable forms of the positives".

Proof of a quarter of the theorem:

First note that:

$x/\mathrm{T} = x \max \mathrm{T};$

$\mathrm{T}/x = x \min \mathrm{T};$

$x/\mathrm{F} = x \min \mathrm{F};$

$\mathrm{F}/x = x \max \mathrm{F}$

and also that for any harmonic f,

$$f(x \min y) \max f(x) = f(x);$$

$$f(x \min y) \max f(x \min(y \max z)) = f(x \min(y \max z)).$$

Therefore:

$$f(a \wedge b) = f(\mathrm{M}(a, \mathrm{F}, b))$$

$$= f((a \min \mathrm{F}) \max (b \min \mathrm{F}) \max (a \min b))$$

$$= f((a/\mathrm{F}) \max (b/\mathrm{F}) \max (a \min b))$$

$$= f(a/\mathrm{F}) \max f((a/\mathrm{F})/(b/\mathrm{F})) \max f((a/\mathrm{F})/(a \min b))$$

$$\max f((b/\mathrm{F})/(a/\mathrm{F})) \max f(b/\mathrm{F}) \max f((b/\mathrm{F})/(a \min b))$$

$$\max f((a \min b)/(a/\mathrm{F})) \max f((a \min b)/(b/\mathrm{F}))$$

$$\max f(a \min b)$$

$$= f(a/\mathrm{F}) \max f(a/\mathrm{F}) \max f(((a \min \mathrm{F})/(a \min b)))$$

$$\max f(b/\mathrm{F}) \max f(b/\mathrm{F}) \max f(((b \min \mathrm{F})/(a \min b)))$$

$$\max f((a \min b)/\mathrm{F}) \max f((a \min b)/\mathrm{F}) \max f(a \min b)$$

$$= f(a \min F) \max f(a \min F/b)$$

$$\max f(b \min F) \max f(b \min F/a)$$

$$\max f(a \min b \min F) \max f(a \min b \min F) \max f(a \min b)$$

$$= f(a \min F) \max f(a \min(F \max b))$$

$$\max f(b \min F) \max f(b \min(F \max a)) \max f(a \min b)$$

$$= f(a \min(F \max b)) \max f(b \min(F \max a))$$

$$= f(a \min(F/b)) \max f(b \min(F/a))$$

$$= f(a) \min f(a/(F/b)) \min f((F/b)/a) \min f(F/b)$$

$$\max$$

$$f(b) \min f(b/(F/a)) \min f((F/a)/b) \min f(F/a)$$

$$= f(a) \min f(a/b) \min f(F/a) \min f(F/(a/b))$$

$$\max$$

$$f(b) \min f(b/a) \min f(F/b) \min f(F/(b/a))$$

$$= (f(a) \min f(a/b) \min f(a \max F) \min f(a/b \max F))$$

$$\max$$

$$(f(b) \min f(b/a) \min f(b \max F) \min f(b/a \max F))$$

$$= (f(a) \min f(a/b)) \max(f(b) \min f(b/a)).$$

The other three-quarters of the theorem are gotten by symmetrical means. If we juxtapose the formulas then we get:

$$f(a \wedge b) = (f(a) \vee f(a/b)/f(b/a)) \wedge (f(b) \vee f(b/a)/f(a/b))$$
$$= (f(a) \wedge f(b/a)/f(a/b)) \vee (f(b) \wedge f(a/b)/f(b/a))$$

$$f(a \vee b) = (f(a) \vee f(b/a)/f(a/b)) \wedge (f(b) \vee f(a/b)/f(b/a))$$
$$= (f(a) \wedge f(a/b)/f(b/a)) \vee (f(b) \wedge f(b/a)/f(a/b)).$$

Define these *diffraction functions*:

$$f_L(a; b) = f(a/b)/f(b/a);$$
$$f_R(a; b) = f(b/a)/f(a/b).$$

Then we get these *diffracted distribution rules*:

$$f(a \wedge b) = (f(a) \vee f_L(a; b)) \wedge (f(b) \vee f_R(a; b))$$
$$= (f(a) \wedge f_R(a; b)) \vee (f(b) \wedge f_L(a; b))$$

$$f(a \vee b) = (f(a) \vee f_R(a; b)) \wedge (f(b) \vee f_L(a; b))$$
$$= (f(a) \wedge f_L(a; b)) \vee (f(b) \wedge f_R(a; b)).$$

Theorem. *Quadrature of majority*:

$$f(\mathrm{M}(a, b, c)) = f((a \min b) \max (b \min c) \max (c \min a))$$

$$= f(a \min b) \max f(b \min c) \max f(c \min a) \max$$

$$f(a \min b/c) \max f(a \min c/b)$$

$$\max f(b \min a/c) \max$$

$$f(b \min c/a) \max f(c \min a/b) \max f(c \min b/a)$$

$$= (f(a) \min f(b) \min f(a/b) \min f(b/a))$$

$$\max (f(b) \min f(c) \min f(b/c) \min f(c/b))$$

$$\max (f(c) \min f(a) \min f(c/a) \min f(a/c))$$

$$\max (f(a) \min f(b/c) \min f(a/c) \min f(b/a))$$

$$\max (f(a) \min f(c/b) \min f(a/b) \min f(c/a))$$

$$\max (f(b) \min f(a/c) \min f(b/c) \min f(a/b))$$

$$\max (f(b) \min f(c/a) \min f(b/a) \min f(c/b))$$

$$\max (f(c) \min f(a/b) \min f(c/b) \min f(a/c))$$

$$\max (f(c) \min f(b/a) \min f(c/a) \min f(b/c)).$$

Similarly with min and max swapped.

This is the "permeable form" of majority.

Corollary. *If f is a positive function,*

$$f(\mathrm{M}(a, b, c)) = \mathrm{M}(f(a), f(b), f(c))$$

$$(because\ f(x/y) = f(x)/f(y)\ for\ f\ positive).$$

B.   Diffraction

a.k.a. harmonic analysis

Recall these *diffraction functions*:

$$f_L(a; b) = f(a/b)/f(b/a);$$
$$f_R(a; b) = f(b/a)/f(a/b) = f_L(b; a).$$

They obey these rules:

$$f_L(a; b)/f_R(a; b) = f(a/b);$$
$$f_R(a; b)/f_L(a; b) = f(b/a);$$
$$f_L(a/b; b/a) = f(a)/f(b);$$
$$f_R(a/b; b/a) = f(b)/f(a).$$

Recall that l and r are the two analytic "side projection" operators;

$$l(x) = x/{}^*x; \quad r(x) = {}^*x/x;$$
$$x = lx/rx; \quad {}^*x = rx/lx;$$

lx is always boolean; so is rx.

Then: $l(f(x)) = f_L(lx; rx);$
$$r(f(x)) = f_R(lx; rx);$$
$$f(x) = f_L(lx; rx)/f_R(lx; rx).$$

f_L and f_R display diamond's typical phase-weaving: if f is a positive function, then:

$$f_L(x; y) = f(x); \quad f_R(x; y) = f(y)$$

if $f = (\sim g)$, then:

$$f_L(x; y) = \sim g_R(x; y); \quad f_R(x; y) = \sim g_L(x; y)$$

so positives preserve phase while negation reverses it.

We can use these rules iteratively to define f_L and f_R "syntactically".

For instance:

If $f(x, y) = x \text{ xor } y = (x \wedge \sim y) \vee (y \wedge \sim x)$, then

$$f_L(x_L, y_L; x_R, y_R) = (x_L \wedge (\sim y_R)) \vee (y_L \wedge (\sim x_R));$$

$$f_R(x_L, y_L; x_R, y_R) = (x_R \wedge (\sim y_L)) \vee (y_R \wedge (\sim x_L)).$$

The sides of x iff y are:

$$\text{iff}_L = (x_L \vee (\sim y_R)) \wedge (y_L \vee (\sim x_R));$$

$$\text{iff}_R = (x_R \vee (\sim y_L)) \wedge (y_R \vee (\sim x_L)).$$

Diffraction can be defined by the *intermix* function:

$$J(a, b) = (a/b, b/a).$$

Note: $J(J(a, b)) = (a, b)$

$$\sim J(\sim a, \sim b) = (b, a);$$

$$^*J(^*a, ^*b) = (b, a)$$

and similarly with the rotation operators.

In dual-rail wiring, J is a simple shuffle gate:

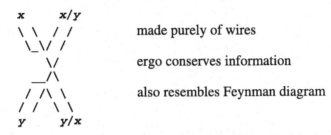

```
x       x/y
 \ \  / /        made purely of wires
  \_\/ /
    \/           ergo conserves information
   _/\
   / /\ \        also resembles Feynman diagram
  / /  \ \
 y      y/x
```

J is its own inverse; therefore we can dualize with J, thus:

$$J \circ (f, f) \circ J(a, b) = (f_L(a; b); f_R(a; b)).$$

Phase separation:

```
a      f     f (a;b)
                L
 \ / \ /
  J   J
 / \ / \
b      f     f (a;b)
                R
```

Dual to this, by conjugation algebra, is:

$$J \circ (f_L, f_R) \circ J(a, b) = (f(a), f(b)).$$

Phase recombination:

```
a    f (a;b)      f(a)
      L
 \ /          \ /
  J            J
 / \          / \
b    f (a;b)      f(b)
      R
```

Phase separation resembles a 2-slit diffraction experiment, with J as the half-silvered mirror, and f as the filter. Similarly, phase recombination resembles a hologram, with phase data reshuffled to retrieve local data.

If we make a phase-separation circuit with two functions f and g, we get $(f/g)_L$ and $(g/f)_R$:

```
    a    f    (f/g)ₗ(a;b)
     \ / \ /
      J   J
     / \ / \
    b    g    (g/f)ᵣ(a;b)
```

These recombine to get f and g back:

```
    a    (f/g)ₗ(a;b)      f(a)
     \ /                   \ /
      J                     J
     / \                   / \
    b    (g/f)ᵣ(a;b)      g(b)
```

J defines these "Diffraction Circuits":

$$d_L(x; y) = x \wedge (\sim y); \quad d_R(x; y) = y \wedge (\sim x)$$

```
    x    d    (x ∧ ~y)
     \ / \ /
      J   J               (and their "∨" = x xor y)
     / \ / \
    y    d    (y ∧ ~x)
```

$$a = x/y; \quad c = da; \quad e = c/d = x - y;$$

$$b = y/x; \quad d = db; \quad f = d/c = y - x;$$

$$g = e \vee f = c \vee d = x \text{ xor } y.$$

Dual to this is:

```
x      (a ∧ ~b)        dx
 \  /              \  /
   J                  J
 /  \              /  \
y      (b ∧~a)        dy
```

(and their "∨" = x/y xor y/x)

$a = x/y$; $c = a \wedge \sim b$; $e = c/d = \mathrm{d}x$;

$b = y/x$; $d = b \wedge \sim a$; $f = d/c = \mathrm{d}y$;

$g = e \vee f = c \vee d = x/y$ xor y/x.

Thus "lower differential is dual to difference".

$D_{\mathrm{L}}(x; y) = y \Rightarrow x$; $D_{\mathrm{R}}(x; y) = x \Rightarrow y$

```
x    D    y⇒x
 \  /  \  /
   J      J
 /  \  /  \
y    D    x⇒y
```

(and their "∧" = x iff y)

$a = x/y$; $c = \mathrm{D}a$; $e = c/d = y \Rightarrow x$;

$b = y/x$; $d = \mathrm{D}b$; $f = d/c = x \Rightarrow y$;

$g = e \wedge f = c \wedge d = x$ iff y.

Dual to this is:

```
x      (b⇒a)      Dx
 \ /        \ /
  J          J        ( and their "∧" =  x/y  iff  y/x)
 / \        / \
y      (a⇒b)      Dy
```

$$a = x/y; \quad c = (b \Rightarrow a); \quad e = c/d = \mathrm{D}x;$$

$$b = y/x; \quad d = (a \Rightarrow b); \quad f = d/c = \mathrm{D}y;$$

$$g = e \wedge f = c \wedge d = (x/y \text{ iff } y/x).$$

Thus "upper differential is dual to implication".

$$\mathrm{D_L}(f; x) = \mathrm{D_R}(x; f) = \sim x;$$

$$\mathrm{d_L}(t; x) = \mathrm{d_R}(x; t) = \sim x$$

```
f    D    ~x              t    d    ~x
 \ / \ /                   \ / \ /
  J   J                     J   J
 / \ / \                   / \ / \
x    D    t               x    d    f
```

$$\sim y = \mathrm{D_L}(f; y);$$

$$\mathrm{D_L}(x; \sim y) = x \vee y;$$

$$\mathrm{D_R}(x; \sim y) = \sim x \vee \sim y;$$

$$a = f/y; \; c = \mathrm{D}a; \; e = c/d; \; f = x/e; \; h = \mathrm{D}f; \; m = h/k;$$

$$b = y/f; \; d = \mathrm{D}b; \; g = e/x; \; k = \mathrm{D}g; \; n = k/h;$$

$$e = \sim y; \; m = x \vee y; \; n = (\sim x) \vee (\sim y);$$

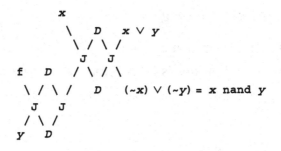

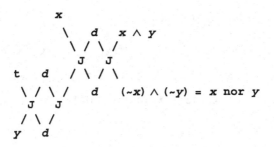

Thus diffraction and differential define all harmonic gates.

Diffraction of the rotations yield both negations:

$L(x/y)/R(y/x) = {}^*y;$

$R(x/y)/L(y/x) = \sim y.$

C.　Buzzers and Toggles

By diffracting a harmonic function's fixedpoints, we get the

Harmonic Self-Analysis Theorem. *For any harmonic* $\mathbf{f}(\mathbf{x})$, *there exists* $\mathbf{x}_L$ *and* $\mathbf{x}_R$, *all of whose components are boolean, such that*

$$(\mathbf{x}_L, \mathbf{x}_R) = J \circ (\mathbf{f}, \mathbf{f}) \circ J(\mathbf{x}_L, \mathbf{x}_R)$$

$$= (\mathbf{f}_L, \mathbf{f}_R)(\mathbf{x}_L, \mathbf{x}_R).$$

That is, $(\mathbf{x}_L, \mathbf{x}_R) = (\mathbf{f}(\mathbf{x}_L/\mathbf{x}_R)/\mathbf{f}(\mathbf{x}_R/\mathbf{x}_L), \mathbf{f}(\mathbf{x}_R/\mathbf{x}_L)/\mathbf{f}(\mathbf{x}_L/\mathbf{x}_R))$.
For any such $\mathbf{x}_L$ and $\mathbf{x}_R$, we have: $\mathbf{f}(\mathbf{x}_L/\mathbf{x}_R) = (\mathbf{x}_L/\mathbf{x}_R)$.
Also, if $\mathbf{f}(\mathbf{x}) = \mathbf{x}$, then $l(\mathbf{x})$ and $r(\mathbf{x})$ fit the above equations.

Thus $\mathbf{x}_L$ and $\mathbf{x}_R$ are the sides of a diamond fixedpoint, themselves forming a twofold fixedpoint. We can find this "diffracted fixedpoint" several ways:

∗ by iteration from $(\mathbf{t}, \mathbf{f})$ or $(\mathbf{f}, \mathbf{t})$;
∗ by $(\wedge, \vee)$ and $(\vee, \wedge)$ on fixedpoints, then iteration;
∗ by (Inf, Cof) and (Cof, Inf) limits of iterations.

Thus by doubling the size of any diamond system, we can harmonically define a boolean analysis of the old system. This is "splitting the circuit". To double is to analyze.

Buzzers-To-Toggle Theorem.

If $(\mathbf{A}, \mathbf{B}) = J(\mathbf{a}, \mathbf{b})$,

i.e. $\mathbf{A} = \mathbf{a}/\mathbf{b}, \mathbf{B} = \mathbf{b}/\mathbf{a}, \mathbf{a} = \mathbf{A}/\mathbf{B},$ *and* $\mathbf{b} = \mathbf{B}/\mathbf{A}$.

Then $\mathbf{f}(\mathbf{a}) = \mathbf{a}$ *and* $\mathbf{f}(\mathbf{b}) = \mathbf{b}$

if and only if

$$(\mathbf{A}, \mathbf{B}) = (\mathbf{f}_L(\mathbf{A}; \mathbf{B}), \mathbf{f}_R(\mathbf{A}; \mathbf{B})) = J[\mathbf{f}, \mathbf{f}](\mathbf{A}, \mathbf{B}).$$

Also, the fixedpoint lattice of the intermixed pair is a junction of two fixedpoint lattices: $L_{J[f,f]} = J(L_f, L_f)$.

Thus two fixedpoints, when intermixed, become an interreferential pair.

The "buzzers-to-toggle" theorem gets its name from this example:

$$(A, B) = J(a, b) :$$

$$a = \sim a; \quad A = \sim a/\sim b = \sim(b/a) = \sim B;$$

$$b = \sim b; \quad B = \sim b/\sim a = \sim(a/b) = \sim A.$$

```
2 lattices: the buzzers        1 lattices: the toggle

                                        tf
                                       /  \
  i  ----  j                         ii    jj
                                       \  /
  i  ----  j                            ft
```

Note that the buzzers, though formally paradoxical, contain boolean information in their phases.

In general, $L_{J[f,f]}$ contains a copy of L_f; its *diagonal*. It also contains a boolean side-analysis of L_f; its *equator*.

For instance, in the toggle lattice, the diagonal {ii, jj} is the lattice for $a = \sim a$; and the equator {tf, ft} are the sides for that lattice.

If we "split the duck":

i.e. if $(A, B) = J(a, b)$, where $a = da$, $b = db$;

then we get the "rabbit":

$A = (A \wedge \sim B);$ $B = (B \wedge \sim A).$

The "Duck" has a linear lattice:

i - - - - f - - - - - j

Therefore the "Rabbit" has a grid-like lattice:

$$\begin{array}{ccccc} & & tf & & \\ & if & & jf & \\ ii & & ff & & jj \\ & fi & & fj & \\ & & ft & & \end{array}$$

The diagonal is {ii, ff, jj}, and the boolean equator is {tf, ff, ft}; duplicates and analysis of the Duck.

The "duck" circuit, when split, yields the "rabbit" circuit; but the "rabbit", when split, just yields two "rabbits". In general any diffracted circuit duplicates when split; that is the "Toggles-to-toggles theorem":

Toggles-To-Toggles Theorem.

If

$$A = f_L(A; a) = f(A/a)/f(a/A);$$
$$a = f_R(A; a) = f(a/A)/f(A/a)$$

and

$$B = f_L(B; b) = f(B/b)/f(b/B);$$
$$b = f_R(B; b) = f(b/B)/f(B/b);$$

then

$$A/B = f_L(A/B; b/a);$$
$$b/a = f_R(A/B; b/a)$$

and

$$B/A = f_L(B/A; a/b)$$
$$a/b = f_R(B/A; a/b).$$

Proof is by direct computation, using the Polarity identity:

$$(u/v)/(x/y) = (u/y).$$

D.   Analytic Diffraction

Recall the **Type Theorem.**

If F does not preserve order, then F is analytic.

Or, to be more specific:

If $F(a) \not\preceq F(b)$ for some $a \preceq b$

then at least one of these two functions equals $- (x)$:

$$F1 = F(a\max(b\min \lambda(x)))/\sim F(a\max(b\min \rho(x)));$$

$$F2 = \sim F(a\max(b\min \lambda(x)))/F(a\max(b\min \rho(x))).$$

The functions $F1$ and $F2$ are derived from diffraction:

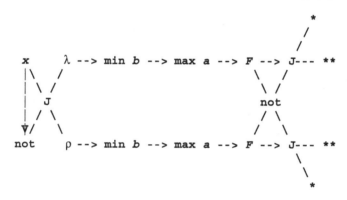

* At least one of these equals $- (x)$;

the other equals one of $\{i, j, x, -(x)\}$

** At least one of these equals $*(x)$;

the other equals one of $\{i, j, \sim x, *(x)\}$.

If we let $G(x) = F(a \max(b \min x))$, then the functions are

$$F1 = G(\lambda(x))/{\sim}G(\rho(x));$$
$$F2 = {\sim}\, G(\lambda(x))/G(\rho(x))$$

and therefore

$$F1 = G_L(x; {\sim}x)/{\sim}G_L({\sim}x; x)$$
$$F2 = {\sim}G_L({\sim}x; x)/G_L(x; {\sim}x).$$

Diffraction is sensitive enough to separate function phases, detect the sides of harmonic fixedpoints, intertwine and untangle fixedpoints, extract minus from analytic functions, and modulate diamond's two reflection operators; yet it is itself defined harmonically, without the use of minus.

Diffraction is harmonic analysis.

E. Diffracting "Two Ducks in a Box"

Recall this Brownian form; "two ducks in a box":

$$C = [[a[a]]_a [b[b]]_b c]_c$$

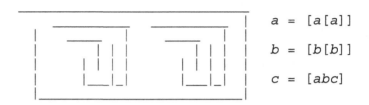

$$a = [a[a]]$$
$$b = [b[b]]$$
$$c = [abc]$$

If we diffract *a* and *b* against each other:

$$A = a/b; \quad B = b/a; \quad C = c$$

then we get this system; "Rabbit in a Box":

$$A = [[A]B];$$
$$B = [[B]A];$$
$$C = [ABC].$$

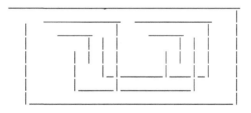

If these are nor gates, then the system is:

$A = A - B =$ I am honest and B is a liar.

$B = B - A =$ I am honest and A is a liar.

$C = \sim(A \vee B \vee C) =$ We are all liars.

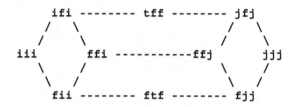

If these are nand gates, then the system is:

$A = B \Rightarrow A =$ If B is honest then so am I.

$B = A \Rightarrow B =$ If A is honest then so am I.

$C = \sim(A \wedge B \wedge C) =$ We are not all honest.

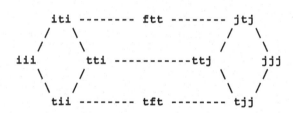

A and B form a "rabbit" circuit; therefore the name "rabbit in a box". C's self-reference splits the void–void state of the rabbit in two; so in 'nor' gates, FF becomes FFI and FFJ; and in 'nand' gates, TT becomes TTI and TTJ.

Chapter 12

How to Count to Two

Brownian and Kauffman Modulators
Diffracting the Modulators
Rotors, Pumps and Tapes
The Ganglion

A. Brownian and Kauffman Modulators

Recall the "**First Brownian Modulator**":

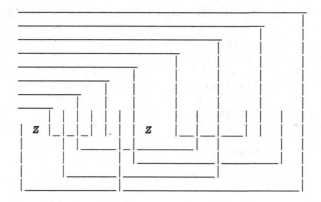

It is equivalent to the boundary-logic system:

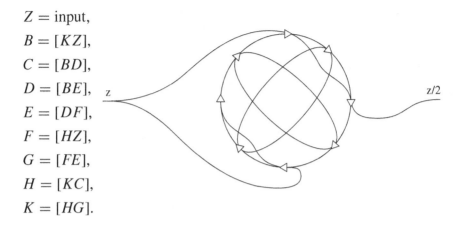

$Z = $ input,
$B = [KZ]$,
$C = [BD]$,
$D = [BE]$,
$E = [DF]$,
$F = [HZ]$,
$G = [FE]$,
$H = [KC]$,
$K = [HG]$.

If we symbolize the marked state by "1", curl by "i", uncurl by "j" and unmarked by "0", then this system has these fixedpoints:

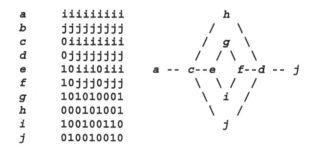

```
a    iiiiiiiii
b    jjjjjjjjj
c    0iiiiiiii
d    0jjjjjjjj
e    10iii0iii
f    10jjj0jjj
g    101010001
h    000101001
i    100100110
j    010010010
```

```
              h
            /   \
           /  g  \
          /  / \  \
a -- c--e    f--d -- j
          \  \ / /
           \  i /
            \  /
             j
```

G. S. Brown, in his *Laws of Form*, claims that this circuit "counts to two"; i.e. when A oscillates twice between marked and unmarked, K oscillates once. Soon we will "diffract" this circuit, and reveal a simple rotor.

Here is a "**Naive Modulator**":

$a = [hz]$,

$b = [gz]$,

$c = [ad]$,

$d = [bc]$,

$e = [c[z]]$,

$f = [d[z]]$,

$g = [eh]$,

$h = [fg]$.

And here is the "**Kauffman Modulator**":

$a = [bjz]$,

$b = [aiz]$,

$c = [ad]$,

$d = [bc] = z/2$;   the half-period oscillator,

$i = [bd]$,

$j = [ac]$.

B.   Diffracting the Modulators

Now we shall use diffraction to analyze these modulators.
Start with the Naive Modulator:

$a = [hz],$

$b = [gz],$

$c = [ad],$

$d = [bc],$

$e = [c[z]],$

$f = [d[z]],$

$g = [eh],$

$h = [fg].$

Let $(A, B) = J(a, b)$; so that $A = a/b$ and $B = b/a$.
Also let $(C, D) = J(c, d)$; $(E, F) = J(e, f)$; $(G, H) = J(g, h)$.
Then:

$A = [Gz],$

$B = [Hz],$

$C = [CB],$

$D = [DA],$

$E = [C[z]],$

$F = [D[z]],$

$G = [GE],$

$H = [HF],$

$G/H = z/2.$

We can write this in a cycle:

$A = [Gz]$,
$D = [AD]$,
$F = [D[z]]$,
$H = [FH]$,
$B = [Hz]$,
$C = [BC]$,
$E = [C[z]]$,
$G = [EG]$,
$z/2 = G/H$.

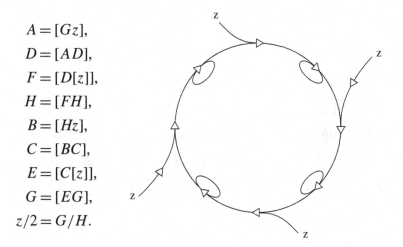

z	A	D	F	H	B	C	E	G	G/H
1	0	i	i	i	0	j	j	j	0
0	j	j	0	i	i	i	0	j	0
1	0	j	j	j	0	i	i	i	1
0	i	i	0	j	j	j	0	i	1

Now try the First Brownian Modulator:

$a = [zd]$,

$b = [ag]$,

$c = [db] = z/2$,

$d = [cf]$,

$e = [cz]$,

$f = [eh]$,

$g = [ah]$,

$h = [eg]$.

Let $(A, E) = J(a, e)$; so that $A = a/e$, and $E = e/a$.

Also let $(B, F) = J(b, f)$; $(C, D) = J(c, d)$; $(G, H) = J(g, h)$.

Then:

$A = [Cz]$,

$E = [Dz]$,

$B = [EH]$,

$F = [AG]$,

$C = [CF]$,

$D = [DB]$,

$G = [GE]$,

$H = [HA]$,

$z/2 = C/D$.

Written in a cycle, you get a "Brownian Rotor":

$$A = [Cz],$$
$$H = [AH],$$
$$B = [HE],$$
$$D = [BD],$$
$$E = [Dz],$$
$$G = [EG],$$
$$F = [GA],$$
$$C = [FC],$$
$$z/2 = C/D.$$

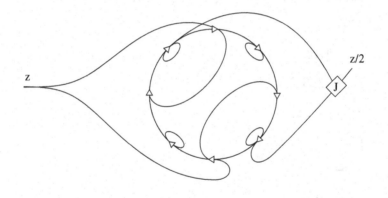

z	A	H	B	D	E	G	F	C	C/D
1	0	i	i	i	0	j	j	j	0
0	j	j	0	i	i	i	0	j	0
1	0	j	j	j	0	i	i	i	1
0	i	i	0	j	j	j	0	i	1

Now try Kauffman's Modulator:

$a = [bdz],$
$b = [ae],$
$c = [df],$
$d = [acz],$
$e = [af] = z/2,$
$f = [de].$

Let $(A, D) = J(a, d)$; so that $A = a/d$, and $D = d/a$.
Also let $(B, C) = J(b, c)$; $(E, F) = J(e, f)$.
Then:

$$A = [ACz],$$

$$D = [DBz],$$

$$B = [DF],$$

$$C = [AE],$$

$$E = [DE],$$

$$F = [AF],$$

$$z/2 = E/F.$$

Written in a cycle, you get a "Kauffman Rotor":

$A = [CAz],$

$F = [AF],$

$B = [FD],$

$D = [BDz],$

$E = [DE],$

$C = [EA],$

$z/2 = E/F.$

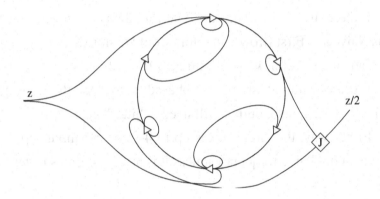

z	A	F	B	D	E	C	E/F
1	0	i	i	0	j	j	0
0	j	j	0	i	i	0	1
1	0	j	j	0	i	i	1
0	i	i	0	j	j	0	0

C. Rotors, Pumps and Tapes

Note that these modulators, once diffracted, become *rotors!* The i's and j's chase each other around a circle.

All these circuits require correct initialization; they need both i's and j's to have "traction". I therefore suggest adding special initialization leads into these circuits.

Note how the states are self-sealing; even when z is off, the i's and j's set up roadblocks against each other. This is due to Duality: [ij] = 0. These circuits demonstrate the purpose of the Second Paradox: to encode phase information.

If these diamond circuits are dual-rail, then each gate is double. Thus Naive and First Brownian rotors would require $2*8 = 16$ gates, and Kauffman's rotor would require $2*6 = 12$ gates.

If these diamond circuits are phased-delay, then the gates must be rigidly timed; also, output will need a "jitter" gate.

Either way, the circuit takes up to 4 steps per input flip. Ergo circuit cannot halve frequency unless it is at least 4 times faster than input.

The "rotor" modulators suggest this system:

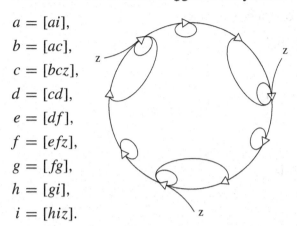

$a = [ai]$,
$b = [ac]$,
$c = [bcz]$,
$d = [cd]$,
$e = [df]$,
$f = [efz]$,
$g = [fg]$,
$h = [gi]$,
$i = [hiz]$.

It's a Kauffman rotor with a third "goldfish" in the loop. This system, when initialized correctly, has period 6:

z	a	b	c	d	e	f	g	h	i
1	i	i	0	i	i	0	j	j	0
0	j	0	i	i	i	i	i	0	j
1	j	j	0	i	i	0	i	i	0
0	i	0	j	j	0	i	i	i	i
1	i	i	0	j	j	0	i	i	0
0	i	i	i	i	0	j	j	0	i

The j's travel around the loop when z toggles.

If we lay out the Brownian or Kauffman rotors out in a line, then we get "pumps", or "one-way tapes":

Brownian pump:

$$\vdots$$
$$A_n = [C_n D_{n-1}],$$
$$B_n = [A_n B_n],$$
$$C_n = [B_n Z],$$
$$D_n = [C_n D_n],$$
$$\vdots$$

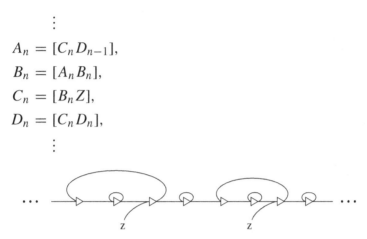

Kauffman pump:

$$\vdots$$
$$A_n = [A_n C_{n-1}],$$
$$B_n = [A_n C_n],$$
$$C_n = [B_n C_n Z],$$
$$\vdots$$

In both of these circuits, the i's and j's travel one block down the one-way tape when Z toggles twice.

By adding a *d*, for direction, variable, we can make "two-way tapes":

Brownian tape:

$$\vdots$$

$A_n = [A_n E_{n-1} D_{n+1}],$
$B_n = [A_n D_n E_n],$
$C_n = [B_n C_n],$
$D_n = [C_n [d] Z],$
$E_n = [C_n dZ],$

$$\vdots$$

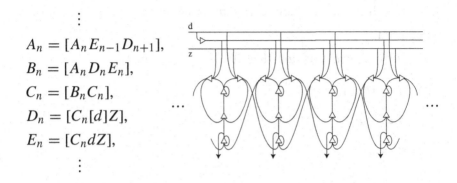

Kauffman tape:

$$\vdots$$

$A_n = [A_n D_{n-1} C_{n+1}],$
$B_n = [A_n D_n C_n],$
$C_n = [B_n C_n [d] Z],$
$D_n = [D_n B_n dZ],$

$$\vdots$$

In these circuits, the i's and j's travel one block along the tape, in the direction indicated by *d*, when *Z* toggles twice. If you connect such a tape to a finite-state machine, then you can make a Turing machine.

D. The Ganglion

Section 2E discussed "phased delay" and "dual rail" implementations of diamond logic. The Ganglion runs on a "phased rail" system, which alternates between evaluating left and right sides of a fixedpoint.

For instance, consider Kauffman's modulator:

$a = [bdz],$

$b = [ae],$

$c = [df],$

$d = [acz],$

$e = [af] = z/2,$

$f = [de].$

If you diffract this into its two sides, then you get the equations:

$a_L = [b_R d_R z];$ $\quad a_R = [b_L d_L z],$

$b_L = [a_R e_R];$ $\quad b_R = [a_L e_L],$

$c_L = [d_R f_R];$ $\quad c_R = [d_L f_L],$

$d_L = [a_R c_R z];$ $\quad d_R = [a_L c_L z],$

$e_L = [a_R f_R];$ $\quad e_R = [a_L f_L],$

$f_L = [d_R e_R];$ $\quad f_R = [d_L e_L].$

To implement this in circuitry, it suffices to have:

Two sets of toggles, to remember the two sides.

Two multiplexers, to compute the disjunctions $b_R d_R z$, $b_L d_L z$, etc.

Two lines of inverters, to compute $[b_R d_R z]$, $[b_L d_L z]$, etc.

Some wiring, switching, etc., to shuffle the above around.

The idea is to compute the left side, given the right side; then store those results into the left side; then compute the right side from the left side; store those results in the right side, and so on.

There should also be bands of wires into and out of the chip, bearing inputs, outputs, multiplexer settings, other controls, and the clock.

If properly constructed, a ganglion would be a virtual chip, composed of phased-rail virtual diamond-valued logic gates; these virtual gates can be wired in any configuration whatsoever, without need for planarity; and they can be rewired in a fraction of a second.

Here is a (very!) rough schematic of the wiring of a Ganglion.

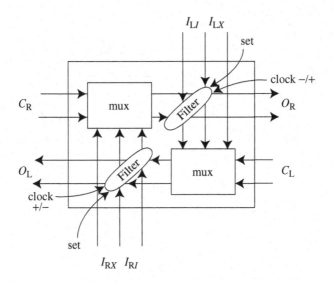

The C_L and C_R inputs control the multiplexer. They could be wired as dual-rail pumps, as above. The O_L and O_R wires are outputs; and I_{LX} and I_{RX} are inputs from outside the ganglion. When the set wire is on, the ganglion's states are reset to I_{LI} and I_{RI}; otherwise the ganglion recycles the O states.

The clock wire alternates $+/-$ or $-/+$. When the clock wire is on, the filter's toggles send data to the next multiplexer; when the clock wire is off, the filter's toggles are reset from the prior multiplexer.

Each filter element has this wiring diagram:

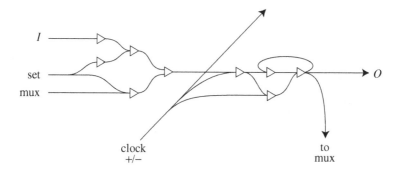

Part Three

Metamathematical Dilemma

Chapter 13

Metamathematics

Gödelian Quanta
Meta-Logic
Dialectic
Dialectical Dilemma

A. Gödelian Quanta

Diamond arises in Gödelian metamathematics. In metamath, sentences can refer to each other's provability, and to quining. This yields self-reference:

T = "'is provable when quined' is provable when quined".

D = "'is unprovable when quined' is unprovable when quined".

S = "'is refutable when quined' is refutable when quined".

P = "'is irrefutable when quined' is irrefutable when quined".

These are self-referential "logic quanta":

T = prv T = the "self-trust" statement,

D = not prv D = the "self-doubt" statement,

S = prv not S = the "self-shame" statement,

P = poss P = the "self-proud" statement,

where prv = provable and poss = possible.

T declares itself believable; it "trusts itself". D declares itself unbelievable; it "doubts itself". S declares itself refutable; it's "ashamed of itself". P declares itself irrefutable; it's "proud of itself".

According to Gödel, D and S are complementary undecidables, corresponding to the paradox-values I and J:

Logic is consistent   if and only if

D is true but unprovable, and S is false but irrefutable.

Logic is inconsistent   if and only if

D is false but provable, and S is true but refutable.

According Löb, T is provably true, and P is provably false; these correspond to the binary truth-values T and F. This is because of the existence of the Gödelian paradoxes; they make the consistency of logic unprovable by any consistent logic; and this collapses the P statement to falsehood.

We get these equations:

$d = \text{poss } T; \quad s = \text{prv } F; \quad p = F; \quad t = T.$

$d = \text{not prv } d = \text{poss not } d = \text{poss } s = \text{poss } T = \text{not prv } F,$

$s = \text{not poss } s = \text{prv not } s = \text{prv } d = \text{prv } F = \text{not poss } T,$

$p = \text{poss } p = \text{not prv not } p = \text{not prv } t = \text{poss } F = F,$

$t = \text{prv } t = \text{not poss not } t = \text{not poss } p = \text{prv } T = T.$

x	not x	prv x	poss x	not prv x	not poss x
t	p	t	d	p	s
d	s	s	d	d	s
s	d	s	d	d	s
p	t	s	p	d	t

To create a diamond logic out of the four Gödelian quanta, we need to identify the "sides". Gödel's analysis provides the key question; namely, is the logic system used consistent or not?

If the system is consistent, then these comparisons apply:

$$P \quad < \quad S \quad < \quad D \quad < \quad T.$$

P is proven false, so it's at the bottom; S is false but not refutable, and so second-from-bottom; D is true but not provably so, and so second-from-top; and T is provably true, so it's at the top.

If the system is inconsistent, then D is false but provable, and S is true but refutable; so the two change places:

$$P \quad < \quad D \quad < \quad S \quad < \quad T.$$

These two order relations agree on the following:

$$
\begin{array}{ccc}
 & D & \\
< & & < \\
P & < & T \\
< & & < \\
 & S &
\end{array}
$$

The minimum and maximum relative to this lattice are diamond's positive functions. So define ∧ and ∨ thus:

Given quanta q and Q,

$q \wedge Q$ = the *biggest* quantum *less* than both q and Q,

in *both* P < S < D < T *and* P < D < S < T orders.

$q \vee Q$ = the *smallest* quantum *more* than both q and Q;

in *both* P < S < D < T *and* P < D < S < T orders.

Note that the minimum of S and D is either S or D, depending on whether the system is consistent; but since we do not know if it is consistent or not, and cannot prove it, it is then prudent to set $S \wedge D = P$, as a concession to our ignorance; and for the same reason we set $S \vee D = T$.

"Not" is boolean: it sends T to P, P to T, D to S, and S to D. Let "*" interchange D and S while leaving T and P fixed. This is the "consistency switch"; it mimics the effect of reversing the consistency of the system. If we define $\sim(x)$ as "not *x*", then we get diamond's negation.

Thus ∧, ∨, and ∼ define a diamond logic for Gödel's quanta.

B.   Meta-Logic

Shame comes to us, proclaiming its own wrongness; it calls itself a liar. If logic is valid, it is wrong, but you cannot be sure. Shame accuses itself; and who are we to contradict it? But by that same token, who are we to believe it? Really it's best to doubt it!

To accuse Shame would be to participate in Shame; why bother? But if we *permit* Shame, then all is well! To admit that self-shame is possible is not self-shame, but self-doubt. What a wonder!

Shame is truly irrefutable in spite of itself. Therefore let there be Shame! Shamelessly allow Shame to exist! Admit its possibility; thus you dispel it!

Doubt comes to us, questioning itself; it calls itself a fool. Yet it is true, if all is well. It comes to us doubting its own good sense; yet if logic is valid, it is honest though dubious. It speaks of a marvel and a wonder, namely itself, and really it's best to wonder at such talk.

Don't believe doubt! Distrust doubt! For the belief in self-doubt is not self-doubt but self-refutation. What a shame!

Thus Doubt is truly dubious. Doubt doubt!

But perhaps it is true anyhow. Even Doubt may be possible.

Pride comes to us, boasting of the great big mathematical model it possesses. But it is wrong! It claims that it will never be refuted; thus it is refuted. Pride is an illusion.

All fantasies proclaim their existence; and they are indeed constructible - on paper. In theory they're practical, but in practice they're theoretical. Pride is just such a fantasy. Pride cannot be wrong; that is its fatal error.

Gödel's Second Incompleteness Theorem describes a truly cosmic catastrophe; for it destabilizes Stability itself. Self-doubt exists, as do self-trust and self-shame, but self-construction does not. Absolute existence does not exist.

Alas! Pride governs markets, churches, states, and empires! Entire civilizations have sold out for Pride's false promises of absolute existence! Pride boasts of its safety, its sanity, its security, its infallibility, its invulnerability, and indeed its immortality! But it is wrong!

Trust comes to us, proclaiming its implicit confidence in itself. Trust believes in Trust; and it is right!

Why? Not because of what it says; but because of what it *doesn't* say: for in fact Trust says nothing at all. It is deductively empty. If you believe Trust, then you believe Trust; that's it. Trust is tautological, i.e. information-free; it proves nothing that was not already provable.

Trust is a vanity. Trust is without content; that's why it's true. Trust can fly because it takes itself lightly. It operates according to the Law of Levity:

Bubbles rise.

What exists?

In other words; what is constructible?

What cannot be refuted?

What is possible?

Pride declares itself possible, but that backfires; whereas the other three quanta are all equally possible:

$$\text{poss } d = \text{poss } s = \text{poss } t = d.$$

poss $s = d$; maybe logic is absurd. Maybe everything is impossible. Maybe nothing exists! I doubt it; but nonetheless the *possibility* that nothing exists does, itself, exist. I personally think it would be a terrible shame if nothing exists; yet on the other hand I think it wonderful that *maybe* nothing exists.

poss $t = d$; maybe trust is possible. Maybe there is self-belief. Maybe necessity exists. In fact I consider that to be an understatement. Necessity doesn't just *exist*; necessity is *necessary*. Vanity is more than just possible; it's universal.

poss $d = d$; maybe doubt is possible. Maybe the unexpected happens. Maybe there is chaos. Maybe wonders exist.

Thus existence, when not outright refutable, reduces to Doubt. What truly exists, then, exists in a state of wonder, a paradox; literally beyond belief. In short, a miracle.

The Doubt statement is a meta-mathematical metaphor for a self-organizing, self-propagating system; its referential twist enables it to transcend stasis. The Doubt statement denotes an organic process whose inherent energy cannot be contained in any fixed stable form. At every stage its paradoxes prevent closure. Doubt is unstable and unreliable; but for that very reason it is irrefutable. It represents Chaos; natural anarchy, which exists if anything does. The anarchy of nature undermines all power systems; arbitrary rules collapse, leaving only those rules which are truly necessary. Thus Natural Anarchy engenders Natural Law; order emerges from chaos.

Because Doubt exists, Trust is valid; from the dynamics of paradox, universal laws derive structure.

Thus cosmos is created by chaos.

Now for an even stranger question:

Do I exist?

What a hazardous question that is! For I am trapped into mathematical error, whether I answer yes or no.

If I deny that I exist, wouldn't that be a self-refutation? What a shame! But worse, if I affirm that I exist, then my affirmation would proclaim its own existence; and that would be quite a thing to be proud of.

For consider; suppose that I were to prove that I exist; and suppose that people duly noted down my proof; and suppose that some scholar came along a century later to inspect my proof. Would it still be valid?

And so I cannot prove that I exist. My existence is truly unprovable; and indeed, it is rather unlikely. When I contemplate the realities of my existence, I perceive that I am a most improbable person. The fact that I exist can only be called a marvel, a wonder, a mystery, and a miracle, utterly beyond belief! And the same goes for you, dear reader; for I doubt that your existence makes any more sense than mine.

Do I exist? I doubt it! So unask the question. The real question is not if I exist (a dubious contingency); it is if I am *necessary*. And my answer to that question is 'yes'. I gladly affirm that — to *me* at least — I am absolutely necessary!

Does that sound vain? Of course it is! Such is the vanity of faith. But that very vanity makes it universal; for don't you, dear reader, consider yourself absolutely necessary — to yourself at least? Don't you have implicit trust in you?

Blessed be the vanity of faith!

C. Dialectic

Now let us consider *pairs* of statements, referring to *each other's* provability. We get this table:

Dialectic

b = a =	prv a	poss a	not prv a	not poss a
prv b	T , T	S , D	S , D	S , S
poss b	D , S	P , P	D , D	D , S
not prv b	D , S	D , D	D , D	P , T
not poss b	S , S	S , D	T , P	S , S

Thus Gödelian dialogs define a 4×4 game. How we score the 16 outcomes is a matter of taste. Generally we will say the "upper outcomes" T and D are worth more than the "lower outcomes" P and S. Below we will use the scoring systems:

S < P < T < D; "soft-edged" scoring;

      the 'soft' outcomes S and D are the extreme values, and

P < S < D < T; "hard-edged" scoring;

      the 'hard' outcomes P and T are the extreme values.

Note the column b = not poss a. Player B has accused player A of being outright refutable. B said to A, "I am *sure* you are wrong!" If you were player A, what would be your best reply?

Clearly it would be to respond a = poss b; that is, to say to player B, "You may be right!" For in that case A equals Doubt and B equals Shame.

Thus "a soft answer turneth away wrath"!

D.   Dialectical Dilemma

Consider these subgames of "Dialectic":

The Doubt-Shame Dilemma (the Gödel game)

b = a =	not prv a	not poss a
not prv b	D , D	P , T
not poss b	T , P	S , S

"Hard-edged" scoring: P < S < D < T.

The Trust-Pride Dilemma (the Löb game)

b = a =	prv a	poss a
prv b	T , T	S , D
poss b	D , S	P , P

"Soft-edged" scoring: S < P < T < D.

Both the "Gödel game" and the "Löb game" are variants of the following game, "Prisoner's Dilemma":

```
a , b                    B
             nice            mean
      ___|_____|_____|
         |       |         |      W  =  Win
 nice    | T , T | L , W   |      T  =  Truce
A        |-------|---------|      D  =  Draw
 mean    | W , L | D , D   |      L  =  Lose
         |_____|_____|
```

scoring: $L < D < T < W$; also $W + L = D + T$

For instance: $(L, D, T, W) = (0, 1, 2, 3)$.

In the "Gödel game", nice = "not prv" and mean = "prv not". In the "Löb game", nice = "prv" and mean = "poss". If players A and B are nice to each other, then they truce, which is better than draw, which they would both get if they were mean to each other; but if only one is nice, that one loses to the other. Both players are tempted to cheat; but if both do, neither wins!

These games set individual gain against mutual gain. No matter what each player says, the other one's best retort is to be mean; but if both do that, then they do worse than if they both are nice. The best shared outcome requires cooperation, but that this cooperation is vulnerable to exploitation.

These Gödelian dialogs reveal meta-logical social dilemmas. Here meta-mathematics meets paradox logic and game theory.

Chapter 14

Dilemma

A.   Milo's Trick

... *"With your permission," said Tock, changing the subject, "we'd like to rescue Rhyme and Reason."*

"Has Azaz agreed to it?" the Mathemagician inquired.

"Yes, sir," the dog assured him.

"THEN I DON'T," he thundered again, "for since they've been banished, we've never agreed on anything — and we never will." He emphasized his last remark with a dark and ominous look.

"Never?" asked Milo, with the slighest touch of disbelief in his voice.

"NEVER!" he repeated. "And if you can prove otherwise, you have my permission to go."

"Well," said Milo, who had thought about this problem very carefully ever since leaving Dictionopolis. "Then with whatever Azaz agrees, you disagree."

"Correct," said the Mathemagician with a tolerant smile.

"And with whatever Azaz disagrees, you agree."

"Also correct," yawned the Mathemagician, nonchalantly cleaning his fingernail with the point of his staff.

"Then each of you agrees that he will disagree with whatever each of you agrees with," said Milo triumphantly; "and if you both disagree with the same thing, then aren't you really in agreement?"

"I'VE BEEN TRICKED!" cried the Mathemagician helplessly, for no matter how he figured, it still came out just that way...

- from *The Phantom Tollbooth*, by Norton Juster

I have long admired the reasoning of this passage, and have sought ways to turn it into mathematics. I have found two ways to do so; by paradox logic, and by dilemma game theory.

By Paradox Logic

Let A = value of utterance made by Azaz, King of Dictionopolis, and let M = truth value of utterance made by the Mathemagician, Wizard of Digitopolis. In the brothers' intractable quarrel, A and M always have opposite values:

$$A = \sim M \quad ; \quad M = \sim A.$$

This system is equivalent to a "toggle" circuit; it can be in one of the two boolean states:

$$A = \text{true}, \ M = \text{false} \quad ; \quad A = \text{false}, \ M = \text{true}.$$

The toggle holds one bit of memory.

But Milo pointed out the symmetry of this system; which implies a symmetrical solution:

$$A = M = \sim A = \sim M.$$

This "singular" solution is a paradox value, equal to its own negation. It is a "null value" of the toggle.

Milo's trick reveals the inner unity from which existence springs.

By Dilemma Game Theory

Here we let A and M argue as before; only this time we let them place a *value* upon the outcome of the argument. This Quarrel game is scored thus:

Win = I'm right, and you're wrong,

Truce = We're both right,

Draw = We're both wrong,

Lose = I'm wrong, and you're right,

where Lose < Draw < Truce < Win.

Truce and win are the "upper" outcomes, or "prosperity"; draw and lose are the "lower" outcomes, or "poverty". Each brother prospers if and only if he is right; and he always gains if the other brother is wrong.

Let Milo ask the brothers if they agree. Each brother can say yes or no; so there are four outcomes:

$$A = \text{``}A = M\text{''} = \text{true} \quad ; \quad M = \text{``}A = M\text{''} = \text{true};$$
$$A = \text{``}A \neq M\text{''} = \text{true} \quad ; \quad M = \text{``}A = M\text{''} = \text{false};$$
$$A = \text{``}A = M\text{''} = \text{false} \quad ; \quad M = \text{``}A \neq M\text{''} = \text{true};$$
$$A = \text{``}A \neq M\text{''} = \text{false} \quad ; \quad M = \text{``}A \neq M\text{''} = \text{false}.$$

Inspection of the table shows that each brother controls the other brother's prosperity. This yields a dilemma game:

(A,M) outcome	M	
	says A = M	says A ≠ M
says A = M	Truce	Lose, Win
says A ≠ M	Win, Lose	Draw

Here, saying $A = M$ is the "nice" move, and saying $A \neq M$ is the "mean" move. The players both prosper only if they *agree to agree*; but this truce is vulnerable to exploitation. One-time play favors draw ("agreeing to disagree"); but in a long tournament, truce can be attained if the players use this strategy:

Do Unto Others As They Have Done Unto You.

That is the "Silver Rule" of reciprocity. Milo's trick guides the brothers to harmony, in the long run.

B. Prisoner's Dilemma

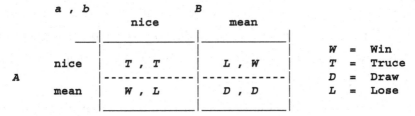

scoring: $L < D < T < W$; also $W + L = D + T$.

For instance: $(L, D, T, W) = (0, 1, 2, 3)$.

This nonzero-sum game presents a player's paradox. It exemplifies the central dilemma of any society; namely, how to get people to co-operate for mutual benefit, when competitive behavior yields a tactical advantage. Negotiation and reciprocation are possible in Dilemma, unlike in competitive games, where there is never anything to negotiate. Mutual profit gives incentive to mutual aid; but exploitation remains tempting.

There are many different strategies for dilemma play. I call three of them the "Iron", "Gold", and "Silver" rules.

The **Iron** rule is the rule of rigid exploitation, justified in the name of expediency. Players ruled by the Iron rule see that no matter how the other player plays, exploitation always yields an advantage; they jump to the conclusion that no more thought is necessary, and play accordingly. This strategy is usually called "All D" (AD) for "Always Defect".

The **Gold** rule is the policy of absolute altruism. Gold rule players see that a society under Golden rule would be at peace, and thus prevail in the long run; they jump to the conclusion that the long run is already here, and play accordingly. This strategy is usually called "All C" (AC) for "Always Cooperate".

The **Silver** rule is the strategy of reciprocity. Silver players do unto others as those others have done unto them. They see that only exact imitation can ensure that the game's inner logic favors cooperation; they jump to the conclusion that the other player is aware of this, and play accordingly. This strategy is usually called "TFT", for "Tit For Tat", which starts by cooperating and continues by reciprocation.

Thus the Iron, Gold, and Silver rules are, respectively, vicious, vulnerable, and vain. Gold is for prey (or host) species, Iron for predator (or parasite) species, and Silver for social (or symbiotic) species. Gold says, "what's mine is yours"; Iron says, "greed is good"; and Silver says, "value for value".

There exist many dilemma strategies other than the Iron, Gold, and Silver rules. For instance, there is R, for Random play; TF2T, "Tit For Two Tats", which defects only after the other player defects twice in a row; 2TFT (two-tits-for-a-tat); "angry" TFT (TFT starting in an unfriendly state); TFT with occasional "testing" behavior; and TFT with "forgiveness factor", which ocassionally (at random) forgives misbehavior on the other player's part; RTFT, "reverse tit-for-tat", which punishes cooperation and rewards punishment; and

"Pavlov", which is nice on the next round if this round truced or drew, and is mean on the next round if this round won or lost. (That is, Pavlov repeats its present play if it came out truce or win, and switches if it came out draw or loss.)

Which strategy is best? That depends on many factors; the other player's strategy, the expected length of the tournament, and the tactical position of the dilemma game itself. Thus dilemma games have a second level of play; strategic as well as tactical. *How* to play matters as much as *what* to play.

Negotiation, strategy, and tactics intermesh in the following two negotiation agendas; "Axial Play" and "The Generous Offer":

Axial Play: for players at balance.
Tactic; a player limits play to truce-draw "axis".
   The board permits no advantage of one over another.
Strategy; that player threatens draw unless truce.
   Appeal to principle. Firmness against exploitation.

This is tactically soft-line cooperative and strategically hard-line competitive. This is the Justice agenda; soft actions, hard bargaining. It stands on shared principle. Its motto is; "Bribe, threaten, and emulate".

The Generous Offer: for player in position of strength.
Tactic; the player limits play to truce-win "column".
   The board permits no adverse outcome for player.
Strategy; the player offers to share his prosperity.
   Appeal to self-interest. Peace bought and paid for.

This is tactically hard-line competitive and strategically soft-line cooperative. This is the Mercy agenda; hard actions, soft bargaining. It stands on shared privilege. Its motto is; "Make them an offer they can't refuse".

Each agenda requires tactical support (the facts on the board) and strategic negotiation (the offer on the table).

C. Dilemma Games

Here is a dilemma version of a familiar game:

Dilemma Tic-Tac-Toe

The grid # and the letters X and O are the same; but there are three new rules:

* Player X starts first, but not in the center square;
* X and O alternate, until they fill the grid; and
* Truce = both XXX and OOO rows; win/lose = only one sort; draw = neither XXX nor OOO rows.

Here are some sample games. Numbers tell order of moves:

X5	O6	X3
O8	O2	X9
X1	X7	O4

Draw

X5	O6	X3
X7	O2	X9
X1	O8	O4

Truce

O4	X3	O6
O8	O2	X9
X1	X7	X5

X wins

X9	X3	O8
O4	O2	O6
X1	X7	X5

Truce

X5	X9	X7
O6	O2	O8
X1	O4	X3

Truce

O6	X3	X5
X7	O2	X9
O4	X1	O8

O wins

O6	X1	O4
X9	O2	X3
X5	O8	X7

Draw

X9	X1	O4
O8	O2	O6
X5	X7	X3

Truce

X5	X7	X3
O8	O2	X9
X1	O6	O4

X wins

Even if we let X start in the center square, we can still get this truce:

O4	X5	O8
O6	X1	X9
O2	X7	X3

Compare these two games:

X5	O6	X3
O8	O2	X9
X1	X7	O4

Draw

X5	O6	X3
X7	O2	X9
X1	O8	O4

Truce

On the second game's sixth move, O put down O6 in the top-center square (A2), and then told X, "If you block me at C2, I'll block you at B1, and we'll draw. Better to grab the ABC1 file now, and let me get ABC2". X agreed, and they truced. This is classic axial play.

Dilemma Chess

Dilemma chess is chess plus deterrence, with a dilemma payoff matrix. The board, pieces and moves are the same as in regular chess; but the game is allowed to end with mutual checkmate, called truce.

This is the payoff matrix:

payoff for (A, B)	B	
	checkmated	not checkmated
A checkmated	(truce, truce)	(lose, win)
not checkmated	(win, lose)	(draw, draw)

Thus a dilemma. If competitive chess is the king of games, then dilemma chess is the queen; for truce opens up a new dimension of play; namely, between competition and cooperation.

The basic innovation in dilemma chess is to allow the "reply" move. The reply move is a final move by the player whose king has been captured. If the other king can be captured in the reply move, then the first capture is "deterred". You may not capture if your check is deterred. You may move into deterred check, or respond to check with a deterrent. You may not cancel the other player's deterrent unless you also escape check (no "forced exchanges").

Mutual deterred check is "tryst"; both sides can capture and retaliate. Truce is mutual assured check (MAC), or inescapable tryst; one move after truce, both sides can still capture and retaliate. In tryst, capture is deterred; the other player could capture next move, but would suffer retaliation. Other forms of deterrence exist; "pinned check", "delayed deterrent", even "temporary checkmate". If you have a deterrent, then your king is free to advance into enemy territory; the "brave king" phenomenon.

For instance, consider this endgame:

	a	b	c	d	e	f	g	h
8							BR	
7	BP			BQ	BP	BP		
6		BP		BP			BK	
5			BP	WP	BB	WP		
4			WP				WK	
3							WP	
2	WP	WP			WN			
1						WR		WQ

Black to move. Note that P × K is deterred.

...		Kg6 × f5	tryst
Qh1-f3	**tryst**	**Kf5-f4**	**tryst**
Kg4-h5	**check**	**Qd7-f5**	**tryst**
Kh5-g6	**tryst**	**Rg8-h8**	**truce**

Two courageous kings!

D.   Dilemma Diamond

The word "truce" rhymes with "true" for good reason. Dilemma games have a four-valued logic, thus:

$$T = T/T = \text{Truce}$$
"true but true"

$$I = T/F = \text{Win/Lose} \qquad\qquad J = F/T = \text{Lose/Win}$$
"true but false" "false but true"

$$F = F/F = \text{Draw}$$
"false but false"

This treats the payoff of a dilemma game as a diamond value. Each dilemma game's value is a pair of Boolean values, thus:

Value of dilemma game G

= (Left gets an upper outcome)/

(Right gets an upper outcome)

= "Left prospers but Right prospers",

where truce and win are the upper outcomes, Left and Right are the two players, and to "prosper" means to truce or win.

Given three dilemma games G, H and K, we define the positive functions "and", "or", "min", "max" and "majority" these ways;

$(G$ and $H)$ = (Left prospers in G and H) /
(Right prospers in G and H)

$(G$ or $H)$ = (Left prospers in G or H) /
(Right prospers in G or H)

$(G$ min $H)$ = (Left prospers in G or H) /
(Right prospers in G and H)

$(G$ max $H)$ = (Left prospers in G and H) /
(Right prospers in G or H)

$M(G, H, K)$ = (Left prospers in most of G, H and K) /
(Right prospers in most of G, H and K)

Recall that diamond has orthogonal reflections, along the horizontal and vertical axes. Horizontal reflection is "not"; vertical reflection is "star".

not (A/B) = (not B)/(not A), $*(A/B) = (*B) / (*A)$.

Through these the two sides interact. Star is the exchange operator; it switches the payoffs for the two players. Not exchanges with a reversal; thus each side gets the opposite of the other's; "we won because they lost".

Considered in dilemma game terms, "not" corresponds to "strategic" distinction. This form of thinking considers the good of the whole; it distinguishes truce and draw, and regards each victory with indifference. "Star" corresponds to "tactical" distinction; this form of thinking considers the good of the individual player; it distinguishes win and lose and regards zero-difference play with indifference.

This matches the two incompatible ways of thinking about Dilemma with the two dual reflections of the logic diamond.

Let Dilemma be given this scoring rule:

Lose $= 0$, Draw $= 1$, Truce $= 2$, Win $= 3$.

If we use the "right side" identification:

Lose $= i$, Draw $= f$, Truce $= t$, Win $= j$,

then dilemma scoring defines a function N:

$$N(i) = 0, \quad N(f) = 1, \quad N(t) = 2, \quad N(j) = 3.$$

This function is a "norm"; that is, it obeys these rules:

$N(x) \geq 0;$

if $x \preceq y$ and $x \neq y$, then $N(x) < N(y)$

$N(x \min y) + N(x \max y) = N(x) + N(y).$

From the norm N, we can define a "metric" D:

$D(x, y) = N(x \max y) - N(x \min y).$

$D(x, y)$	I	F	T	J
I	0	1	2	3
F	1	0	3	2
T	2	3	0	1
J	3	2	1	0

It is a theorem of lattice theory that any such metric D defined from a norm N has these properties:

$D(x, y) \geq 0$ for all x and y;

$D(x, y) = 0$ if and only if $x = y$;

$D(x, y) = D(y, x)$;

$D(x, z) \leq D(x, y) + D(y, z)$.

The dilemma metric defines "distance" in diamond. Thus dilemma makes diamond not just a logic, but a *space* as well.

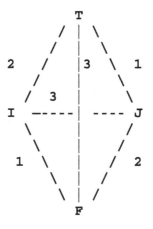

E.　Banker's Dilemma

A Billiard-Marker, whose skill was immense
might perhaps have won more than his share;
But a Banker, engaged at enormous expense
had the whole of their cash in his care.

— Lewis Carroll, *The Hunting Of The Snark*

Consider a dilemma game between players A and B; it is financed by a banker C, who gets to keep the remainder of the fund after the payoffs are distributed. Their payoffs are:

4 dollars invested

```
(A,B,C) payoff              B
                   nice            mean
        -----|---------------|---------------|
             |               |               |
   nice      |   (2,2,0)     |   (0,3,1)     |
             |               |               |
A       -----|---------------|---------------|
             |               |               |
   mean      |   (3,0,1)     |   (1,1,2)     |
             |_____|_____|
```

This makes dilemma zero-sum again; for the main player's cooperation is the banker's defeat. What is more, the banker has a vested interest in fostering distrust between the other two players. (Indeed, that is the *only* thing the banker can actively do; for the other two players make all the moves.)

The three players rank win/lose, lose/win, truce, and draw in three different ways:

$$A: \quad L/W \quad < \quad D \quad < \quad T \quad < \quad W/L$$
$$B: \quad W/L \quad < \quad D \quad < \quad T \quad < \quad L/W$$
$$C: \quad T \quad < \quad W/L \quad = \quad L/W \quad < \quad D$$

These three preference rankings yield these majorities:

2/3 say:	$W/L \quad < \quad D$	(Voters B and C)
2/3 say:	$L/W \quad < \quad D$	(Voters A and C)
2/3 say:	$D \quad < \quad T$	(Voters A and B)
2/3 say:	$T \quad < \quad W/L$	(Voters A and C)
2/3 say:	$T \quad < \quad L/W$	(Voters B and C)

W/L $\qquad\qquad\qquad\qquad\qquad\qquad\qquad$ L/W

$\qquad <_{BC}$ $\qquad\qquad\qquad\qquad$ $<_{BC}$

$\qquad\qquad$ $D \quad <_{AB} \quad T$

$\qquad <_{AC}$ $\qquad\qquad\qquad\qquad$ $<_{AC}$

L/W $\qquad\qquad\qquad\qquad\qquad\qquad\qquad$ W/L

This "Condorcet Crossing" diagram agrees with most–but not all–of each voter's preferences. It contains preference loops, yet every player agrees that the order relation is transitive! Thus we get a voter's paradox. The banker's financing makes dilemma zero-sum, but non-Aristotelian. The glitch remains; to escalate order is to escalate chaos.

Either non-modus-ponens or nonzero-sum; Dilemma's illogic is marked. It often displays paradoxical signs, for Dilemma is, so to speak, snarked.

F. The Unexpected Departure

For truce to succeed requires certain conditions. One of them is that the expected number of plays be great enough; another is that the play not end at too definite at time. If it does, then a "backwards induction paradox" destroys truce, no matter how long the tournament.

Consider the following scene:

Curly is about to play with Moe in a dilemma tournament scheduled to last exactly 100 rounds. Curly, a Silver Rule player, is optimistic that he can convince Moe (an Iron Rule player) that it'll be in his own best interest to cooperate.

But Moe said, "What about the 100th round? Won't that be the last"?

Curly said, "Yes".

"There won't be any rounds after the 100th"?

"Yes", said Curly.

Moe asked, "So in the very last play, what's to keep me from defecting"?

"Cause then I'll defect the next..." Curly said, then slapped himself on the face. "Alright, nothing will stop you from defecting on the 100th play".

"So you might as well defect too, right"? Moe said, smiling.

"I guess so", Curly said reluctantly. "On the 100th play".

Moe continued, "And what about the 99th play? What's to keep me from defecting then"?

"Cause then I'll defect the next..." Curly said, then slapped himself on the face. "But I'll defect on the 100th play anyhow".

"That's right", Moe said, smiling.

"So nothing's keeping you from defecting on the 99th play".

"That's right", Moe said, smiling.

"So I should defect on the 99th play also", said Curly.

"That's right", Moe said. "Now, what about the 98th play"?

And so they continued! Moe whittled down Curly's proposed truce, one play at a time, starting from the end. By the time the conversation was over, Moe had convinced Curly that the only logical course was for them to defect from each other 100 times, drawing the tournament. And so they did; yet when Curly played with Larry (a Gold Rule player) they cooperated 100 times, for a truce!

Thus we deduce, by mathematical induction, that the prospect of abruptly terminated play, even if in the far future, poisons the relationship at its inception.

That is the "backwards induction paradox". In dilemma play, cooperation requires continuity to the end. Departure should not be at an expected time lest that light the backwards-induction fuse; departure should be unannounced, at an unexpected time.

We need an *unexpected* departure; but this yields a paradox. Consider this following story about an Unexpected Exam:

Once upon a time a professor told his students, "Sometime next week I will give you an exam; and that exam will be at an unexpected time. Right up until the moment I give you the exam, you will have no way to deduce when it will happen, or even if it will happen. It will be an Unexpected Exam".

One of the professor's student objected, "But then the exam couldn't happen on Friday; for by then it would be expected"!

The professor said, "True".

The student continued, "So Friday's ruled out".

Another student said, "But if Thursday's the last possible day for an Unexpected Exam, then it's ruled out too; for by Thursday the *Thursday* exam will be expected"!

The professor said, "True".

And so on; by such steps the students concluded that the Unexpected Exam can't happen on Friday, Thursday, Wednesday, Tuesday, or Monday; so it can't happen at all!

"So you don't expect it"? said the professor.

His students said, "No"!

The professor smiled . . .

On the next Wednesday, he handed out an exam, to his students' surprise.

That's the Paradox of the Unexpected Exam. Here a backwards induction paradox also appears; but this time it yields a strangely *false* result rather than a strangely *undesirable* result. This match of methods suggest the following fable.

The same professor visited the Dean; he said, "I will depart this school sometime during the next month. To ensure cordial relations between us until that time, my departure will take place on an unexpected day. It will be an Unexpected Departure".

The Dean retorted, "You couldn't leave on the 31st, for by then your Unexpected Departure would be expected".

The professor agreed.

The Dean added, "Having ruled out the 31st, the 30th is also ruled out; for *it* would be expected".

The professor agreed to that too.

And so the conversation continued; and in the end the Dean concluded, "Your Unexpected Departure can't happen on any day. Therefore I don't expect it". The professor agreed.

On the seventeenth day of the month the professor departed, to the Dean's astonishment.

This **Paradox of the Unexpected Departure** is just what the doctor ordered; for here the *failure* of backwards induction (so puzzling to the reason) is precisely what is needed to defend the Axelrod equilibrium from *its* backwards induction proof!

A dilemma tournament can use "open bounding"; replay only if a random device permits it. This ensures an Unexpected Departure; play will be finite, but there will be no definite last play during which the Iron player is safe from the danger of Silver retaliation.

The paradox of the Unexpected Departure is related to the paradox of the First Boring Number; for presumably the tournament ends as soon as it stops being interesting.

The conclusion then is clear; let none of your social relationships end too definitely; let there be some possibility that you might encounter that person again, soon. (And conversely, when you *must* leave, slip away quietly!)

Notes

Chaos on the boundary; four interpretations; differentials; categoricity?; a parenthetical remark; complementarity santa and grinch; lattice emulation?; contra well-foundedness; permeable forms and direct distribution; double diffraction; other modulators; Tertullian quanta; diamond values for dilemma?

1E.   Chaos on the Boundary

George Spencer-Brown, in his book *Laws Of Form*, noted that you can, if you wish, derive a Liar-like paradox from the square root of minus one.

For if i equals the square root of -1, then it solves the equation $x = -1/x$. But you get into trouble if you try to solve this equation in the real numbers. If x is positive then $-1/x$ is negative; if x is negative then $-1/x$ is positive; if x is zero then $-1/x$ is infinite; if x is infinite then $-1/x$ is zero.

If you iterate $-1/x$ then you get the oscillating sequence

$$x, -1/x, x, -1/x, x, -1/x, \ldots.$$

These are slopes of perpendicular lines; i as a 90-degree rotation. We can extract square roots by Newton's Method:

$$x_{n+1} = (x_n + A/x_n)/2 = (x_n^2 + A)/(2x_n).$$

For positive A, this converges rapidly to the square root of A. But if A is negative one, then you get:

$$x_{n+1} = (x_n - 1/x_n)/2 = (x_n^2 - 1)/(2x_n).$$

Note this angle-doubling trigonometric identity:

$$\cot(2\theta) = (\cot^2(\theta) - 1)/(2\cot(\theta)).$$

So if $x_0 = \cot(\theta)$, then $x_1 = \cot(2\theta)$; and in general $x_n = \cot(2^n\theta)$.

This angle-doubling iteration is chaotic; that is, it has sensitive dependence on initial conditions, topological mixing, and dense periodic orbits.

If x_0 has a small imaginary part, then the iteration converges quickly to i or to $-i$. The real line is the boundary between two basins of attraction.

So when you inspect the sign of i, you get oscillation; a sign of paradox. And when you solve for i by Newton's Method, you get chaos on the boundary; another sign of paradox.

2F.   Interpretations

There are four interpretations of boundary logic in glut-gap logic:

[]	= true	; true	; false	; false
6	= gap	; glut	; gap	; glut
9	= glut	; gap	; glut	; gap
[[]]	= false	; false	; true	; true
[ab]	= a nor b	; a nor b	; a nand b	; a nand b

These are isomorphic to each other under conjugation by four operations: identity, swap TF, swap GG, swap both pairs - a Klein 4-group.

3B. Differentials

One can rewrite the Primary Normal Form in terms of differentials:

$$F(x) = B_1(x) \vee (B_2 \wedge dx) = b_1(x) \wedge (b_2 \vee Dx),$$

where no B_i and no b_i have any dx or Dx terms. This is the "derivative series" form. In two variables it's:

$$F(x, y) = B_1(x, y) \vee (B_2(y) \wedge dx) \vee (B_3(x) \wedge dy) \vee (B_4 \wedge dx \wedge dy)$$
$$= b_1(x, y) \wedge (b_2(y) \vee Dx) \wedge (b_3(x) \vee Dy) \wedge (b_4 \vee Dx \vee Dy)$$

– and so on, for more variables, using higher order differentials.

The differentials have these laws:

$$dx \wedge dy = (x \wedge y) - (x \vee y) \text{ "both without either"}$$
$$= d(x/y) \wedge d(y/x) \text{ "intermix"}$$

$$dx \vee dy = (dx \text{ iff } dy) \wedge (dx \text{ iff } Dy)$$
$$= (x \text{ iff } y) \wedge (x \text{ iff } {\sim}y)$$
$$= (x/y) \text{ xor } (y/x)$$

$$Dx \vee Dy = (x \wedge y) \Rightarrow (x \vee y) \text{ "both implies either"}$$
$$= D(x/y) \vee D(y/x) \text{ "intermix"}$$

$$Dx \wedge Dy = (Dx \text{ xor } Dy) \vee (Dx \text{ xor } dy)$$
$$= (x \text{ xor } y) \wedge (x \text{ xor } {\sim}y) \text{ "opposite reflections"}$$
$$= (x/y) \text{ iff } (y/x)$$

In Boolean logic, "xor" and "iff" are isomorphic to addition modulo 2. Alas, in diamond they are no longer group operations:

$$(t \wedge i) \text{ xor } (t \wedge i) = i \neq f = (t \text{ xor } t) \wedge i,$$
$$(f \vee i) \text{ iff } (f \vee i) = i \neq t = (f \text{ iff } f) \vee i$$

so they are non-distributive;

$$(t \text{ xor } i) \text{ xor } j = t \neq f = t \text{ xor } (i \text{ xor } j)$$
$$(f \text{ iff } i) \text{ iff } j = t \neq f = f \text{ iff } (i \text{ iff } j)$$

so they are non-associative;

$$\text{not}(i \text{ xor } j) = t \neq f = (\text{not } i) \text{ xor } j$$
$$\text{not}(i \text{ iff } j) = f \neq t = (\text{not } i) \text{ iff } j$$

so they are non-symmetrical.

This is because xor and iff contain asymmetric differential terms:

$$x \text{ iff } y = \sim(x \text{ xor } y) = (\sim x \text{ xor } y) \vee \mathrm{d}x \vee \mathrm{d}y$$
$$x \text{ xor } y = \sim(x \text{ iff } y) = (\sim x \text{ iff } y) \wedge \mathrm{D}x \wedge \mathrm{D}y$$

$$(x \text{ and } z) \text{ xor } (y \text{ and } z) = ((x \text{ xor } y) \wedge z) \vee ((x \vee y) \wedge \mathrm{d}z)$$

$$(x \vee z) \text{ iff } (y \vee z) = ((x \text{ iff } y) \vee z) \wedge ((x \wedge y) \vee \mathrm{D}z)$$

$$(x \text{ xor } (y \text{ xor } z)) = (x \wedge y \wedge z) \vee (\sim x \wedge \sim y \wedge z) \vee (\sim x \wedge y \wedge \sim z)$$

$$\vee (x \wedge \sim y \wedge \sim z) \vee (x \wedge \mathrm{d}y) \vee (x \wedge \mathrm{d}z)$$

$$(x \text{ iff } (y \text{ iff } z)) = (\sim x \vee \sim y \vee \sim z) \wedge (\sim x \vee y \vee z) \wedge (x \vee \sim y \vee z)$$

$$\wedge (x \vee y \vee \sim z) \wedge (\sim x \vee \mathrm{D}y) \wedge (\sim x \vee \mathrm{D}z)$$

3C. Categoricity?

Conjecture: Diamond is a "categorical" DeMorgan algebra:

That is, any De Morgan algebra is isomorphic to a subalgebra of products of copies of diamond. These De Morgan algebras need not have the Complementarity axiom; they are isomorphic to subalgebras of ones that do.

If this conjecture is true, then diamond is to De Morgan algebras as two-valued logic is to boolean algebra. I consider diamond to be a two-dimensional extension of two-valued logic that solves paradox, just as the complex plane is a two-dimensional extension of the real line that solves $x^2 = -1$.

4A. A Parenthetical Remark about the Parenthetical Remark

Consider the Brownian form [[]]. It is equal to (i.e. confused with) the void; yet it is not itself void, being made of two nested marks. It therefore deserves names of its own; I suggest "doublecross", or the "remark". Doublecross denotes the void, but unlike the void, is visible.

The remark is to forms as zero is to numbers; both name the nameless, and both are placeholders. In algebraic terms, the remark denotes parentheses:

$$(A) \quad = \quad \overline{\overline{A \mid} \mid} \quad = \quad A$$

I use parentheses to distinguish these from the brackets of boundary logic. In fact (A) = [[A]].

The remark allows one to express the associative law:

$$(AB)\,C \;=\; \overline{\overline{A\ B}\ |}\ |\ C \;=\; ABC \;=\; A\ \overline{\overline{B\ C}\ |}\ |\ =\; A(BC)$$

[[A]] *remarks* about A without *marking* A; it draws attention without changing values. It is literally a parenthetical remark.

Now consider this *re-entrant remark*:

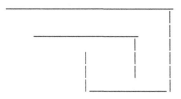

It represents this system:

$$A \;=\; \overline{B}\ | \qquad ;$$

$$B \;=\; \overline{A}\ | \qquad .$$

$$A \;=\; (\,A\,) \qquad .$$

This is a *toggle*, or *memory circuit*.

Thus memory remarks on itself.

5D. Complementarity Santa and Grinch

Recall the Brownian form; "two ducks in a box":

$$C = [[a[a]]_a[b[b]]_b \, c]_c$$

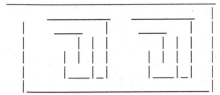

$$a = [a[a]]$$
$$b = [b[b]]$$
$$c = [abc]$$

If these are nand gates, then:

a	$=$	Da	$=$	"I am honest or a liar".
b	$=$	Db	$=$	"I am honest or a liar".
c	$=$	$\sim(a \wedge b \wedge c)$	$=$	"One of us is a liar".

I call this the "Complementarity Santa", because sentence c is of the form

$$c = (c \Rightarrow (da \vee db))$$

– a Santa sentence trying to make lower differentials disjoin to true.

But if these are nor gates, then:

a	$=$	da	$=$	"I am honest and a liar".
b	$=$	db	$=$	"I am honest and a liar".
c	$=$	$\sim(a \vee b \vee c)$	$=$	"All of us are liars".

I call this the "Complementarity Grinch", because c is of the form:

$$c \;=\; ((Db \wedge Db) - c)$$

– a Grinch sentence trying to make upper differentials conjoin to false.

It has this fixedpoint lattice:

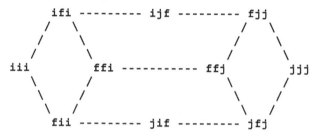

Note the fixedpoints ijf and jif; these are the only one where C has a boolean value; but this is due to Complementarity, an anti-boolean axiom.

Without C, this lattice would be a 3×3 square; but C's negative self-reference splits the center state. The TT state, in nands, splits into TTI and TTJ; the FF state, in nors, splits into FFI and FFJ.

General Lattice Emulation?

Is *every* lattice the fixedpoint lattice for some harmonic system? If so, then given a lattice, how do we find a system that emulates it?

Given a system, how can we simplify it? (Reduce the number of variables, references, etc.)

How does a lattice change when you change its system? And vice versa?

Given a lattice and its system, can input leads into the system provide control over points in the lattice?

What *practical* computation tasks does lattice emulation permit? Could we (say) emulate a *tree* (for sorting, searching, etc.) via a lattice? Note that boolean logic usually corresponds to the *equator* of the lattice, or the *nodes* of the tree; the part with the fewest lattice constraints.

9A.   Contra Well-Foundedness

Set theorists responded to Russell's Paradox by insisting that all sets be "well-founded" – that is, that no infinite descending element chains exist. This rules out all sets that contain themselves; therefore all sets are assumed to be elements of Russell's set. Well-foundedness therefore makes Russell's set equal to the universal set; but since the existence of Russell's set is in dispute, this seems to me not the best way to found set theory.

Why insist that no set contains itself? Why not insist instead that *every* set contains itself? That's just as logical, and just as irrelevant. Well-foundedness treats the symptom, and not the disease, which is boolean thinking. Accept paradox logic, and Russell's Paradox loses all of its force.

And what is a well-founded set? It is a collection; that is, a predicate, a property; and it applies to other well-founded sets. A well-founded set is a property of properties of properties, etc., and the chain always ends with the empty set. And what is the empty set? It too is a property; the one that applies to no set at all. The empty set is the property of untruth.

So a so-called "well-founded" set is a property of properties, etc., always ending in nullity. It's an iterated description of *falsehood*.

I don't think that's so well founded. In fact I think that's the worst possible foundation. It's like building a castle, not on sand, but on vacuum.

Even the term *foundation* is misleading. It says that the set world is like a tower, resting heavily upon solid ground; but what supports the ground? Physics and astronauts assure us that "down" is relative, that the Earth beneath our feet is floating free in space, resting upon itself.

11A.   Permeable Forms and Direct Distribution

Any function $g(\mathbf{x})$ can be put in disjunctive normal form:

$$g(\mathbf{x}) = \text{OR[all } i](s_{i1}(x_1) \wedge \cdots \wedge, s_{in}(x_n)),$$

where each $s(x)$ equals t, or x, or $\sim x$, or dx.

Use this fact, along with the Phased Distribution rules:

$$f(a \wedge b) = (f(a) \min f(a/b)) \max(f(b) \min f(b/a))$$
$$= (f(a) \max f(b/a)) \min(f(b) \max f(a/b)),$$

$$f(a \vee b) = (f(a) \min f(b/a)) \max(f(b) \min f(a/b))$$
$$= (f(a) \max f(a/b)) \min(f(b) \max f(b/a)).$$

Proceeding by iteration, we can distribute any f over g, until we get a lattice operation over the set $\{f(\pm x_i / \pm x_j)\}$, where $\pm x_i$ is either (x_i) or it is $(\sim x_i)$.

Thus we derive this theorem:

Permeable Form Theorem. *For any* $g(\mathbf{x})$, *there exists a lattice function* L_g *(made from* $\min$, $\max$) *such that* $g(\mathbf{x}) = L_g(\pm x_i / \pm x_j)$; *$g$'s "permeable form".*

Furthermore, for any harmonic f,

$$f(g(\mathbf{x})) = L_g(f(\pm x_i / \pm x_j)).$$

This is **General Quadrature**, or **semi-distribution**.

For instance, here are the permeable forms for d, D, $\Rightarrow$, and minus:

$$f(dx) = f(x \wedge \sim x)$$
$$= (f(x) \min f(x/\sim x)) \max (f(\sim x) \min f(\sim x/x))$$
$$= (f(x) \max f(\sim x/x)) \min (f(\sim x) \max f(x/\sim x)).$$

$$f(Dx) = f(x \vee \sim x)$$
$$= (f(x) \max f(x/\sim x)) \min (f(\sim x) \max f(\sim x/x))$$
$$= (f(x) \min f(\sim x/x)) \max (f(\sim x) \min f(x/\sim x)).$$

$$f(x \Rightarrow y) = f(\sim x \vee y)$$
$$= (f(\sim x) \max f(\sim x/y)) \min (f(y) \max f(y/\sim x))$$
$$= (f(\sim x) \min f(y/\sim x)) \max (f(y) \min f(\sim x/y)).$$

$$f(x - y) = f(x \wedge \sim y)$$
$$= (f(x) \min f(x/\sim y)) \max (f(\sim y) \min f(\sim y/x))$$
$$= (f(x) \max f(\sim y/x)) \min (f(\sim y) \max f(x/\sim y)).$$

Here's an open question: when can we reduce Quadrature to two terms, yielding **direct distribution**:

$$f(x \min y) = f(x) \min f(y); \quad f(x \max y) = f(x) \max f(y).$$

For functions of more than one variable, direct distribution is over each variable separately:

$$f(x \min y, z) = f(x, z) \min f(y, z);$$

$$f(x, y \min z) = f(x, y) \min f(x, z)$$

and similarly for max.

The distributive functions in one variable include the constant functions t, f, the identity function, and the negation function $\sim x$; but *not* the differentials:

$$d(t \min f) = di = i \quad ; \quad dt \min df = f \min f = f;$$

$$D(t \min f) = Di = i \quad ; \quad Dt \min Df = t \min t = t.$$

The distributive functions in two variables include:

the constant functions t, f;

the junction function x/y;

$x \wedge y$;

$x \vee y$;

$x - y$;

$x \Rightarrow y$;

x nand y;

x nor y;

– but *not* the diagonal functions (x xor y) or (x iff y):

j iff (t min f) = j iff i = t; (j iff t) min(j iff f) = j min j = j,

j xor (t min f) = j xor i = f; (j xor t) min(j xor f) = j min j = j.

Conjecture. The **Anchored Lattice Forms**.

A function f is distributive if and only if it is identical to both of its anchored lattice forms. In one variable these forms are:

$$f(x) = (\lambda(x) \min f(\text{F})) \max(\rho(x) \min f(\text{T})) \max f(\text{I}),$$

$$f(x) = (\rho(x) \max f(\text{F})) \min(\lambda(x) \max f(\text{T})) \min f(\text{J}).$$

In two variables the anchored lattice forms are:

$$f(x, y) = (\lambda x \min \lambda y \min f(\text{F}, \text{F}))$$

$$\max(\lambda x \min \rho y \min f(\text{F}, \text{T})),$$

$$\max(\rho x \min \lambda y \min f(\text{T}, \text{F}))$$

$$\max(\rho x \min \rho y \min f(\text{T}, \text{T})) \max(\lambda y \min f(\text{I}, \text{F}))$$

$$\max(\rho y \min f(\text{I}, \text{T})) \max(\lambda x \min f(\text{F}, \text{I}))$$

$$\max(\rho x \min f(\text{T}, \text{I})) \max(f(\text{I}, \text{I}))$$

$$f(x, y) = (\lambda x \max \lambda y \max f(\text{T}, \text{T}))$$

$$\min(\lambda x \max \rho y \max f(\text{T}, \text{F}))$$

$$\min(\rho x \max \lambda y \max f(\text{F}, \text{T}))$$

$$\min(\rho x \max \rho y \max f(\text{F}, \text{F})) \min(\lambda y \max f(\text{J}, \text{T}))$$

$$\min(\rho y \max f(\text{J}, F)) \min(\lambda x \max f(\text{T}, \text{J}))$$

$$\min(\rho x \max f(\text{F}, \text{J})) \min(f(\text{J}, \text{J})).$$

Three variables and up have "hypercubical" versions of these, with anchor terms at corners, edges, faces, solids and up.

Define a **diagonal operator** as one of these:

$$x \wedge \sim x \qquad\qquad ; \qquad x \vee \sim x$$

$$(x \vee y) \wedge (\sim x \vee \sim y) \qquad ; \qquad (x \wedge y) \vee (\sim x \wedge \sim y)$$

$$(x \vee y \vee z) \wedge (\sim x \vee \sim y \vee \sim z) \;;\; (x \wedge y \wedge z) \vee (\sim x \wedge \sim y \wedge \sim z)$$

$$(x \vee y \vee z \vee w) \qquad ; \qquad (x \wedge y \wedge z \wedge w)$$

$$\wedge (\sim x \vee \sim y \vee \sim z \vee \sim w) \qquad \vee (\sim x \wedge \sim y \wedge \sim z \wedge \sim w)$$

and so on. I call the functions in the first column *"sumbunol"* functions, for "some but not all"; and I call the functions in the second column *"nunnerol"* functions, for "none or all". Denote these as "sbnl" and "nnrl".

Theorem. Differential Junction Square

$$\text{sbnl}(x_1, x_2) = d(x_1/x_1) \vee d(x_1/x_2) \vee d(x_2/x_1) \vee d(x_2/x_2),$$

$$\text{nnrl}(x_1, x_2) = D(x_1/x_1) \wedge D(x_1/x_2) \wedge D(x_2/x_1) \wedge D(x_2/x_2).$$

And in general:

$$\text{sbnl}(x_1, \ldots, x_N) = \text{OR}_{i, j \leq N}\, d(x_i/x_j),$$

$$\text{nnrl}(x_1, \ldots, x_N) = \text{AND}_{i, j \leq N}\, D(x_i/x_j).$$

Conjecture. Diagonal Extraction

If F is not distributive, then there exist distributive functions f_i such that: $F(f_1(x_1), f_2(x_2), f_3(x_3), \ldots)$ equals a diagonal operator.

10D.   Double Diffraction

The Buzzers-To-Toggle Theorem can be generalized to two functions:

If $(\mathbf{A}, \mathbf{B}) = J(\mathbf{a}, \mathbf{b})$ then:

$\mathbf{f}(\mathbf{a}) = \mathbf{a}$ and $\mathbf{g}(\mathbf{b}) = \mathbf{b}$

if and only if

$(\mathbf{A}, \mathbf{B}) = ((\mathbf{f}/\mathbf{g})_L(\mathbf{A}; \mathbf{B}), (\mathbf{g}/\mathbf{f})_R(\mathbf{A}; \mathbf{B}))$

```
 A     f     (f/g) (A;B)
  \ / \ /         L
   J   J
  / \ / \
 B    g     (g/f) (A;B)
                 R
```

```
 a    (f/g)       f(a)
  \ /      L   \ /
   J           J
  / \         / \
 b    (g/f)      g(b)
          R
```

If $a = \mathrm{d}a$; $b = \mathrm{D}b,$
and $(A, B) = J(a, b),$
then: $A = A \max {\sim} B$; $B = B \min {\sim} A.$

If $a = \lambda(a)$; $b = \rho(b)$
and $(A, B) = J(a, b),$
then: $A = A$; $B = {\sim} A.$

A is free, and B is its negation.

12B. Other Modulators

Consider this form; the "Second Brownian Modulator":

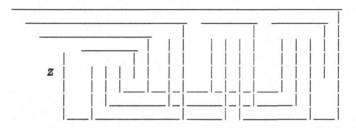

It is equivalent to this system:

$$Z = \text{input},$$
$$A = [ZBE],$$
$$B = [ZAC],$$
$$C = [ABD],$$
$$D = [BCF] = Z/2,$$
$$E = [ABC],$$
$$F = [ACD].$$

Let $(a, b) = J(A, B)$, so that $a = A/B$, $b = B/A$.
Also let $(c, e) = J(C, E)$ and $(d, f) = J(D, F)$.
Then:

$$a = [z\,a\,c],$$
$$b = [z\,b\,e],$$
$$c = [a\,b[[C/D]]] = [a\,b[[c/f]]],$$
$$e = [a\,b[[D/C]]] = [a\,b[[d/e]]],$$
$$d = [a\,C\,d] = [a[[c/e]]d],$$
$$f = [b\,C\,f] = [b[[c/e]]f],$$
$$z/2 = d/f.$$

z	a	b	c	e	d	f	d/f
1	0	0	j	i	i	j	1
0	j	i	0	0	j	i	0
1	0	0	i	j	0	0	0
0	i	j	0	0	i	j	1

We can simplify the diffracted form slightly:

$$a = [z\,a\,c],$$
$$b = [z\,b\,e],$$
$$c = [a\,b\,c\,f],$$
$$e = [a\,b\,d\,e],$$
$$d = [a\,C\,d] = [a[[c/e]]d],$$
$$f = [b\,C\,f] = [b[[c/e]]f].$$

Here's the "Interval New Reductor":

$$Z = \text{input},$$
$$A = [BD] = Z/2,$$
$$B = [AZ],$$
$$C = [BDZ],$$
$$D = [CF],$$
$$E = [BF],$$
$$F = [EC].$$

Let $(a, f) = J(A, F)$, so that $a = A/F$, $f = F/A$.

Also let $(b, c) = J(B, C)$ and $(d, e) = J(D, E)$.

Then:

$$a \quad = \quad [c\,e],$$

$$f \quad = \quad [b\,d],$$

$$b \quad = \quad [[[BD/A]]Z] \quad = \quad [[[b\,d/f]]Z],$$

$$c \quad = \quad [[[A/BD]]Z] \quad = \quad [[[a/c\,e]]Z],$$

$$d \quad = \quad [bF] \qquad\qquad = \quad [b[[f/a]]],$$

$$e \quad = \quad [cF] \qquad\qquad = \quad [c[[f/a]]],$$

$$Z/2 \quad = \quad a/f.$$

z	a	f	b	c	d	e	f/a	a/f
1	0	0	0	0	1	1	0	0
0	j	i	i	j	0	0	1	0
1	1	1	0	0	0	0	1	1
0	i	j	j	i	j	i	0	1

Note that a and f rotate around the logic diamond!

13A.   Tertullian Quanta

There are quanta other than the Gödelian quanta. For instance, there is the Tertullian quantum. I name it after Tertullian, who once declared that he believed a certain point of doctrine "because it is impossible". Charmed by so defiant a folly, I decided to recast his little joke into meta-mathematical terms.

Let a "Tertullian" quantum be of the form:

"Believe this statement only if it is absurd".

Trtl $=$ prv(Trtl) $\Rightarrow$ prv($\sim$Trtl).

If you could prove Trtl, then you could prove its opposite; but that would be absurd. Therefore belief in Trtl implies belief in absurdity, as it says; therefore it is true, though you cannot prove it. Therefore doubt it!

Equivalent forms exist. The logical principle "modus tollens" assures us that the statement "$A \Rightarrow B$" equals "not $B \Rightarrow$ not A"; therefore the Tertullian quantum can be rephrased thus:

"This sentence is consistent only if it is unprovable".

"I am possible only if I am incredible".

These formulae seem more modernistically skeptical than Tertullian's aggressive faith, but they are logically equivalent.

Therefore my reply to Tertullian is:

He is irrefutable because he is unbelievable.

"Possibile est quia incredibile est".

It is possible because it is unprovable.

Tolerate me because I am dubious!

14C.   Diamond Values for Dilemma?

Dear reader, I *used* to believe the following conjecture:

Valuation Conjecture: Any dilemma game position's diamond value can be evaluated via iteration to a fixedpoint.

This fixedpoint is found by assigning a diamond value to each node of a "game tree"; the graph describing all game positions and moves. The diamond values can be computed for each node, in terms of the other nodes, following a formula given by the game tree. Thus we get an inter-referential system of diamond values; and any such system has at least one fixedpoint.

I call Valuation a "conjecture" rather than a "theorem" because of certain difficulties in the above proof. Specifically, it seems impossible to capture the entire Dilemma game-outcome valuation by either harmonic or paraharmonic functions. The trouble is that Dilemma evaluates the outcomes in a *linear* (numerical) lattice, but diamond is not such a lattice. Paraharmonic functions capture the zero-sum nature of competitive thought; likewise, harmonic functions capture the zero-difference nature of cooperative thought; but Dilemma transcends both.

The value of the game depends on how two players think about the game; thus, by definition, no single system of reasoning can capture Dilemma's essence. Dilemma requires dialog.

Bibliography

Robert Axelrod, *The Evolution of Cooperation* (Basic Books, 1984).

Nuel D. Belnap, "A Useful 4-valued Logic", in *Modern Uses of Multiple-Valued Logic*, eds. J. M. Dunn and G. Epstein (Reidel, 1977).

William Bricken, "Syntactic Variety in Boundary Logic", Diagrams 2006, LNAI 4045, pp. 73–87 (2006).

W. Bricken and E. Gullichsen, "Introduction to Boundary Logic", *Future Computing Systems* 2(4) (1989) 1–77.

Nicholas Falletta, *The Paradoxicon* (Doubleday & Company, 1983).

Patrick Grim, *The Philosophical Computer* (MIT Press/Bradford Book, 1977).

Anil Gupta, "Truth and Paradox", *J. Philosophical Logic* 11 (1982) 1–82.

Nathaniel Hellerstein, "Diamond, a Four-Valued Approach to the Problem of Paradox", U.C. Berkeley, Doctoral thesis, 1984.

Nathaniel Hellerstein, "Diamond, a Logic of Paradox", Cybernetic, Summer-Fall 1985, v.1.

Nathaniel Hellerstein, *Diamond, A Paradox Logic* (World Scientific, 1997).

Nathaniel Hellerstein, *Delta, A Paradox Logic* (World Scientific, 1997).

Douglas Hofstadter, *Gödel, Escher, Bach* (Basic Books, 1979).

Patrick Hughes and George Brecht, *Vicious Circles and Infinity* (Penguin, 1975).

Norton Juster, *The Phantom Tollbooth* (Random House, 1961).

J. A. Kalman, "Lattices with Involution", *Trans. Amer. Math. Soc.* **87** (1958) 485–491.

Louis Kauffman, "De Morgan Algebras — Completeness and Recursion", in *Proc. of the 8th Int. Symp. on Multiple-Valued Logic* (1978), pp. 209–213.

Louis Kauffman, "Imaginary Values in Mathematical Logic", in *Proc. of the 17th Int. Symp. on Multiple-Valued Logic* (IEEE, 1987).

Louis Kauffman, "Knot Automata", *24th Int. Symp. on Multiple-Valued Logic* (IEEE, 1994).

Louis Kauffman, "On the form of self-reference", in *Systems Inquiry*: *Theory, Philosophy and Methodology*, ed. Bela H. Bathany (Inter-systems Pub., 1985), pp. 206–210.

Louis Kauffman, "Self-Reference and Recursive Forms", *J. Social Biol. Structure* **10** (1987) 53–72.

Louis Kauffman, "Special Relativity and a Calculus of Distinctions", in *Proc. of the 9th Annual Meeting of ANPA*, Cambridge (ANPA West, 1987), pp. 290–311.

Louis Kauffman, "Network Synthesis and Varela's Calculus", *Int. J. Gen. Syst.* **4** (1978) 179–187.

Louis Kauffman and Francisco Varela, "Form Dynamics", *J. Social Biol. Structure* **3** (1980) 171–206.

Saul Kripke, "Outline of the Theory of Truth", *J. Philosophical Logic* **72** (1975) 690–716.

Ernest Nagel and James Newman, *Gödel's Proof* (New York University Press, 1958).

James Newman, *The World of Mathematics* (Simon and Schuster, 1956).

William Poundstone, *Labyrinths of Reason* (Anchor Press, Doubleday, 1988).

Graham Priest, "The Logic of Paradox", *J. Philosophical Logic* **8** (1979) 219–241.

Graham Priest, *In Contradiction* (Martinus Nijhoff, 1987).

Rudy Rucker, *Infinity and the Mind* (Bantam Books, 1982).

Richard Shoup, "A Complex Logic for Computation", Interval Research Corporation.

Raymond Smullyan, *This Book Needs No Title* (Prentice-Hall, 1980).

Raymond Smullyan, *Forever Undecided* (Knopf, 1987).

Raymond Smullyan, *To Mock A Mockingbird* (Knopf, 1985).

George Spencer-Brown, *Laws of Form* (The Julian Press, 1979).

Francisco Varela and J. Goguen, "The Arithmetic of Closure", *J. Cybernetics* **8** (1978).

Francisco Varela, "A Calculus for Self-Reference", *Int. J. Gen. Syst.* **2** (1975) 5–24.

Francisco Varela, "The Extended Calculus of Indications Interpreted as a Three-Valued Logic", *Notre Dame J. Formal Logic* **17** (1976).

Steve Yablo, "Grounding, Dependence and Paradox", *J. Philosophical Logic* **11** (1982) 117–137.

Index

SERIES ON KNOTS AND EVERYTHING

Editor-in-charge: Louis H. Kauffman *(Univ. of Illinois, Chicago)*

The Series on Knots and Everything: is a book series polarized around the theory of knots. Volume 1 in the series is Louis H Kauffman's Knots and Physics.

One purpose of this series is to continue the exploration of many of the themes indicated in Volume 1. These themes reach out beyond knot theory into physics, mathematics, logic, linguistics, philosophy, biology and practical experience. All of these outreaches have relations with knot theory when knot theory is regarded as a pivot or meeting place for apparently separate ideas. Knots act as such a pivotal place. We do not fully understand why this is so. The series represents stages in the exploration of this nexus.

Details of the titles in this series to date give a picture of the enterprise.

*The complete list of the published volumes in the series, can also be found at
http://www.worldscibooks.com/series/skae_series.shtml